Principia Designae – Pre-Design, Design, and Post-Design

AF789485

Toshiharu Taura
Editor

Principia Designae – Pre-Design, Design, and Post-Design

Social Motive for the Highly Advanced Technological Society

Springer

Editor
Toshiharu Taura
Organization of Advanced Science and Technology
Kobe University
Kobe, Japan

ISBN 978-4-431-56341-9 ISBN 978-4-431-54403-6 (eBook)
DOI 10.1007/978-4-431-54403-6
Springer Tokyo Heidelberg New York Dordrecht London

© Springer Japan 2015
Softcover reprint of the hardcover 1st edition 2015
This work is subject to copyright. All rights are reserved by the Publisher, whether the whole or part of the material is concerned, specifically the rights of translation, reprinting, reuse of illustrations, recitation, broadcasting, reproduction on microfilms or in any other physical way, and transmission or information storage and retrieval, electronic adaptation, computer software, or by similar or dissimilar methodology now known or hereafter developed. Exempted from this legal reservation are brief excerpts in connection with reviews or scholarly analysis or material supplied specifically for the purpose of being entered and executed on a computer system, for exclusive use by the purchaser of the work. Duplication of this publication or parts thereof is permitted only under the provisions of the Copyright Law of the Publisher's location, in its current version, and permission for use must always be obtained from Springer. Permissions for use may be obtained through RightsLink at the Copyright Clearance Center. Violations are liable to prosecution under the respective Copyright Law.
The use of general descriptive names, registered names, trademarks, service marks, etc. in this publication does not imply, even in the absence of a specific statement, that such names are exempt from the relevant protective laws and regulations and therefore free for general use.
While the advice and information in this book are believed to be true and accurate at the date of publication, neither the authors nor the editors nor the publisher can accept any legal responsibility for any errors or omissions that may be made. The publisher makes no warranty, express or implied, with respect to the material contained herein.

Printed on acid-free paper

Springer is part of Springer Science+Business Media (www.springer.com)

Preface

Many countries have successfully developed highly technological societies. However, there remain some key issues that cannot be explained within the conventional framework of design research or methodology. In other words, there are some matters that must be captured outside of the conventional design process purview. In particular, social and cultural notions of design should be considered from a technological viewpoint. In this book, a wider and deeper design purview is discussed within the framework of "Pre-Design, Design, and Post-Design" by focusing on the "motive of design."

The conventional design process comprises three stages: conceptual design, embodiment design, and detailed design. In order to capture the inherent nature of design, we believe that it is not sufficient to investigate only these three design stages, but rather it is also necessary to look at not only the design phases that occur before (the Pre-Design phase) and after (the Post-Design phase) the design process but also the relationships between the phases. In the Post-Design phase, consumers use the products, and based on those consumer–product interactions (users experience positive or negative aspects of a product through either successfully using or failing to successfully use a product), consumers' feelings and criteria, or awareness of problems relating to existing or new products are created and stored in society either explicitly or implicitly. These newly formed feelings and criteria, or awareness are called "social motive," which is expected to be a key aspect in discussions around the deeper design purview. In the Pre-Design phase, an explicit or implicit social motive is identified and translated into an explicit requirement or specification for a new product either deductively (e.g., improvement of utility or efficiency), inductively (e.g., marketing survey), or abductively (e.g., creator's intuition).

The Pre-Design phase has been studied within the framework of idea generation, concept generation, marketing surveys, risk-management, etc., whereas the Post-Design phase has been studied within the framework of product usability, emotional design, user-centered design, etc. However, these areas have been approached independently and have not yet been systematized. In particular, little

Fig. 1 Workshop participants in front of the Nara Hotel

attention has been paid to the link from the Post-Design phase to the Pre-Design phase, in which the notion of social motive is expected to play an important role.

To discuss this issue, we held a workshop, "Design Research Leading Workshop—Nara Workshop," on August 24–25, 2013, at the Nara Hotel in Nara, Japan. There were 18 workshop participants: Petra Badke-Schaub (TU Delft, the Netherlands); Hernan Casakin (Ariel University, Israel); Gaetano Cascini (Politecnico di Milano, Italy); Amaresh Chakrabarti (Indian Institute of Science, India); John Gero (George Mason University, USA); Gabriela Goldschmidt (Technion—Israel Institute of Technology, Israel); Noriko Hashimoto (Aoyama Gakuin Woman's Junior College, Japan); Michio Ito (Tokyo University of Agriculture and Technology, Japan); Shuichi Iwata (The Graduate School of Project Design, Japan); Pascal Le Masson (MINES ParisTech, France); Udo Lindemann (Technical University Munich, Germany); Chris McMahon (University of Bristol, UK); Martin Steinert (Norwegian University of Science and Technology, Norway); Barbara Tversky (Stanford University, USA); Yukari Nagai (Japan Advanced Institute of Science and Technology, Japan); Toshiharu Taura (Kobe University, Japan); Georgi Georgiev (Kobe University, Japan); and Kaori Yamada (Kobe University, Japan) (Fig. 1).

In the workshop, we engaged in two types of discourse. The first was a group discussion followed by short presentations from each participant. The second was a group discussion in which a topic was given in advance and participants exchanged ideas about the topic followed by the presentation of some information referring to the topic provided by workshop coordinators. We had two discussions of the latter type:

The first discussion topic was "art and society in design". The preceding presentation on the topic is summarized:

> The presentation focused on obtaining clues about "art and society in design" from Japanese five-story pagoda structures. Particularly, the Kohfukuji pagoda, which is visible from the workshop room, was introduced. The symbolism of pagodas in history and their five-layered shape that symbolically corresponds to human life was explained. The center pillar (heart) of the structure is the main unique characteristic of the Japanese five-story pagoda; the pillar functions as a balance, preventing the structure from falling down in case of an earthquake.

In the discussion, several ideas were presented:

- Every structure (which refers to something constructed, such as a building, pagoda, etc.) has meaning. The assignment of names to such structures creates the different meanings between the structures.
- The transfer of technology from traditional pagodas to contemporary tower (Tokyo Skytree)—which use the same design concept of balance but adapt the concept to new possibilities—is notable. In terms of the influence of culture on technology, the issue that technology inspires construction or vice versa is essential. It is a combination of things—evolved technologies and cultural needs—that come together, influence one another and create outstanding phenomena. The assigning of symbolism comes from somewhere else outside of this combination. Design is a confluence between creating a need and offering a solution that applies to a cultural issue. An invention may not immediately connect to a need.
- Not all cultures automatically generate the same kinds of technologies. The way technologies are used and developed is localized, context specific, time specific, and needs specific. Currently, the intellectual property, proprietary documentation, and knowledge on how to design something stay within individual organizations. It is key to share real design knowledge.
- Constraints are incredibly useful for creativity. There are examples of fast technological developments due to constraints and counterexamples of slow developments. Sometimes non-availability of technology might be cost-effective.
- How to transfer technological knowledge from one generation to the next is embedded and inherent in society. Examples of this knowledge transfer include how to carry heavy Japanese floats (people pull heavy, decorated floats through towns by hand during festivals). Furthermore, in many examples technology expands feelings deep in our mind and contributes to our culture and its improvement. For example, mold-casting technology from 2,000 years ago

allowed people to create beautiful art. The question remains whether the same condition exists now and how we can recognize the relationships between technology and culture. Modern society is full of absolutely transformational technologies (e.g., transistors and software). Lost technologies and the need to consider local culture in technology should be discussed.

The second discussion topic was "science and technology in design". The preceding presentation on the topic is summarized:

> The presentation focused on obtaining clues about "science and technology in design" from nuclear technology. Particularly, the relationship between science and technology, products, and society were the focus of this presentation. The ways in which science, technology, and products are accepted by society were explained: they must be part of culture; the social system must be constructed to cope with any negative aspects; and there is a need to cognitively recognize the phenomena. The presentation included how to cope with certain aspects of technology and how to manage its negative aspects, responsibilities, and risks, such as those related to nuclear technology, so that the product is not isolated from society.

In the discussion, several ideas were presented:

- The notions of risk and human understanding are unrelated; an example of this can be observed in the probability of being killed by a car. The notion of risk is developed on the basis of experience. Therefore, the question is how to develop a new notion of risk. Risk management cannot be directly applied to the case of nuclear power. Risk perception is essential, and we often take bigger risks than when other people subject us to risk. There are tremendous new safety developments within nuclear power plants. We should shut down the old and build new, better, and safer power plants.
- Possible main causes of fatalities are the accidents and mistakes that occur in hospitals. For example, there is a need for a better engineering solution for a joint anesthetic system.
- The hope for the future exists on a global scale. Knowledge on the recognition of and responsibility for human errors must be translated to the next generation. Due to the lack of real data, society is very sensitive to certain technologies, such as airplanes and nuclear power; however, compared to the oil and coal industries, whose related global-warming issues are uncontrollable, nuclear technology global-warming issues are possible to control. We may have the responsibility to prepare society with basic information for decision making about nuclear technology and to show all of the related risks in order for people to have open and knowledgeable discussions.
- The notion of risk is not automatically developed based solely on experience. To establish a notion of risk, not only must data be accepted but a feeling must also be accepted. All parties (not only scientists and nuclear technology specialists but also philosophers) need to be included in this discussion. There exists a great debate on how to control and compare the risks of nuclear technology. Risk is a design paradigm, particularly when establishing whether the variety of generated solutions is sustainable. Some fields have many innovations, while others do not.

- An issue is the general public's acceptance or rejection of novel technologies in terms of why people accept or reject novelties. This is a matter of identifying invisible mechanisms. Resistance to change is an example.

On the basis of the previous discussions, the "General discussion" started with the identification of keywords:

> Belief, Identity, Experience, Expectations, Aspirations, Incentive, Norm, Individual, Collective, Social, Constraints, Meaning, Risk, Sign and Design, Pagoda, Culture, Interdisciplinary, Holism, Systems, Interdependency, Situatedness, Evolution, Context, Philoxenia, Iterability, Independence, Exaptation, Design Legitimacy, Intrinsic and Extrinsic Motivation, Wishes and Proposals, Social–Environmental–Personal, Requirements, Emotions and Experiences, Creativity, Invisible, Ethics, Feeling, Evaluation, Time, Communication, Consumer Behavior, Motives, Responsibility, Social Changes, Transdisciplinary, Adoption, Technology, Flexibility, Resilience, Philosophy, Institution, Research, Sustainability, Politics, Interpretation, Imagination, Metaphor, Analogy, Innovation, Economy, Embeddedness, Network, and Priorities.

After the identification of keywords, several ideas were presented:

- The interaction between design and society exists on various levels. We are referring to different kinds of societies when speaking about individuals (perceived differences), consumers (individuals who belong to cultures), and designers (we are speaking in universal terms).
- The notion of insight is important. The notion of insight deals with particular issues, such as sources of inspiration and constraints of identity and knowledge. Design does not exist without insight.
- The notion of responsibility is notable. Questions on how a designer is given responsibility and how designers can change the world should be discussed. Where and when do designers take responsibility and how to teach them to take responsibility? Designers must have an open mind to consider solutions with different consequences.
- There are a huge number of different cultures and related complexities. Designers live and work in different cultures, companies, and departments. They need to be aware of the network of interdependencies. We need to study more cultures and look for common traits among different cultures through interacting with people from different domains. Creativity is based on the notion of being different from, or how to add to, something already in existence. It is difficult to innovate when ignoring cultural differences.
- The common culture of engineers (also designers, architects) is important. There is a question regarding the difference between design cultures. A common culture can be traced through history, books, education, and professors. Engineers can think in a universal and largely accepted manner. Uniformity exists in the way of tackling certain engineering problems. Each engineer is sensitive to context. Looking at artifacts in cultures is essential. There are strong differences between artifacts and their characteristics within different countries.
- Even though the function of an artifact is the same, due to individual cultures, there may be a completely different understanding of the same design problem.

Different cases that focused on details or concepts can be exemplified. Truly different design cultures can be exemplified, as engineers do not understand each meaning in design. Different representations of design exist: in practice, design is concerned with the artifact, not the function. There are different languages and insufficient communications in academia and in practice, which may contribute to a problem in design.

Based on the discussions in the workshop, the participants contributed to this book their thoughts about the Pre-Design, Design, and Post-Design phases. Rivka Oxman (Technion—Israel Institute of Technology, Israel) also joined the authors. This book consists of 14 chapters. To help readers understand the bigger picture, the 14 chapters have been classified under four parts.

Part I is entitled "General and Philosophical Consideration on Pre-Design, Design, and Post-Design" and contains three chapters. Toshiharu Taura introduces the framework of Pre-Design, Design, and Post-Design followed by an outline of the relationship between technology, products, and society. To capture the technology, he also touches on three views: culture, social system, and cognition. Discussing the characteristics of creativity in engineering design, Udo Lindemann stresses that "there should not be any real Pre-Design phase, as long as we discuss design (product development) cycles in a holistic way and not as a specific project. And similar is the point of Post-Design phase." Furthermore, he introduces a process model that shares similarities with Pre-Design, Design, and Post-Design. Michio Ito captures the relationships between Pre-Design, Design, and Post-Design from the viewpoint of communication between the sender and the receiver of information. He points out that design in general must be able to function in the absence of the sender, the receiver, and the context of production (i.e., they must not lose their function in different settings), and there is always the possibility of transformation (iterability).

Part II is entitled "Culture and Cognition of Pre-Design, Design, and Post-Design" and contains four chapters. Yukari Nagai extracts the social motives of advanced design that represent "sustainability" from a sense of nature, "openness" as a self-organizing system from a sense of culture, and "deep design thinking" for a sense of future throughout the discussion of critical issues with regard to creativity in design in human and social cognition. Gabriela Goldschmidt presents a comprehensive notion of design that includes the Pre-Design and Post-Design phases as hinging on an extended design space that is modeled as comprising two main components: design expertise and creativity. She links the model to the metaphorically indestructible Japanese five-tier pagoda. Hernan Casakin and Shulamith Kreitler introduce a new approach to social and personal motivations for creativity grounded in the cognitive orientation (CO) theory and discuss its relation as a major drive to design motive in the Pre-Design and Post-Design phases. Barbara Tversky discusses the notions of Design and Redesign by focusing on the role of sketches. She argues that just as designers call their interactions with their sketches "conversations," design and redesign can be viewed as conversations between designers and users.

Part III is entitled "Technology and Society of Pre-Design, Design, and Post-Design" and contains three chapters. Shuichi Iwata analyzes the Fukushima nuclear reactor accident in the framework of three loops for learning: how to learn, changing the rules, and following the established rules. He discusses an extension of data-driven materials design by introducing human dimensions for nuclear reactors and adding irreversible path-dependent features and human dimensions for environmental issues. Amaresh Chakrabarti undertakes a preliminary enquiry into the broad processes that might constitute the Pre-Design, Design, and Post-Design phases, and proposes the "design-society" cycle as a framework for further enquiry into these processes, to understand their influence on: developing knowledge and experience triggered in the societal mind by a product, subsequent transformation of this knowledge and experience to form new product requirements, and further transformation of this knowledge and requirements to form new products. Noriko Hashimoto introduces the notion of "negative technology," which was proposed by Tomonobu Imamichi and focuses on the phase after the making of new products (the Post-Design phase in this book). She stresses that "in the twenty-first century, it is necessary to think about negative technology (technica negativa). After Fukushima, we must concern ourselves about safety and close the nuclear facilities. Such a project must be a negation of positive technology, but this negativity should be turned to be positive."

Part IV is entitled "Models of Pre-Design, Design, and Post-Design" and contains four chapters. Russell Thomas and John Gero present a system to model social interactions between producers and consumers in the Post-Design phase. They suggest that the computational experiments enable us to test hypotheses regarding the mutual influence of producer and consumer values on the trajectory of design improvements. Gaetano Cascini and Francesca Montagna propose an extension of the Gero's Function–Behavior–Structure (FBS) framework aimed at representing Needs, Requirements, and their relationships with the Function, Behavior, and Structure of an artifact. They suggest that the extended model may support a more careful and detailed investigation of the processes that occur in the earliest stages of design, and specifically what happens in new product development activities. Pascal Le Masson, Benoit Weil, and Olga Kokshagina discuss a new perspective for risk management by focusing on two questions: Can we design an alternative that would lower risk, and does this new alternative create new risks? They conduct a study of the design of generic technology with a matroid model in the C-K theory. Rivka Oxman presents a computational medium for social knowledge acquisition, design integration, and design adaptation. She describes the potential of, and approaches to, crowdsourcing design by adapting concepts such as the wisdom of the crowd.

This book is just the beginning in our efforts to capture design in a wider and deeper purview. We hope that this book will help readers consider Pre-Design, Design, and Post-Design from the viewpoint of culture, cognition, technology, and society; further discussions will be held.

I am thankful to all the participants in the workshop and to the authors for writing their chapters for this book. We had very intensive discussions in the workshop and these discussions have evolved in this fruitful book. The conceptions of this book are based on ideas I had together with Professor Michio Ito (Tokyo University of Agriculture and Technology, Japan) and Professor Yukari Nagai (Japan Advanced Institute of Science and Technology). I would like to express deep gratitude to them both. In addition, we would like to express our sincere gratitude to Dr. Georgi V. Georgiev for his devoted contributions towards the editing of this book. Finally, we are thankful to the Faculty of Engineering and the Organization of Advanced Science and Technology of Kobe University for their support for the workshop. We also express our thanks to Springer Japan, especially to Dr. Yuko Sumino, for publishing this book.

Toshiharu Taura

Contents

About the Authors

In the order of chapters

Toshiharu Taura is the dean of and a professor at the Organization of Advanced Science and Technology at Kobe University. He is also a professor in the Mechanical Engineering Department. He received his B.S., M.S., and Dr. Eng. degrees from the University of Tokyo, Japan, in 1977, 1979, and 1991, respectively. After serving as a mechanical engineer at the Nippon Steel Corporation and as an associate professor at the University of Tokyo, Taura joined Kobe University in 1999. He founded the Design Creativity Special Interest Group in the Design Society in 2007 and launched the *International Journal of Design Creativity and Innovation* in 2013. He is also a fellow of the Design Research Society.

Udo Lindemann studied Mechanical Engineering and did his Ph.D. in product development and systems engineering.

In the years to follow, he held a number of leading positions in industry.

He took over the Institute of Product Development at the Technical University of Munich in 1995. Today he is the head of the Academic Senate of the Technical University Munich.

He is co-publisher of the German journal "Konstruktion" and co-editor of several international journals. From 2007 to 2010 he served as President of the Design Society. In addition he is an active member of a number of scientific societies and other organizations. In 2008 he became an elected member of the German Academy of Science and Engineering.

Michio Ito is an associate professor at Tokyo University of Agriculture and Technology. He received B.A. and M.A. from Osaka University in philosophy. After engaging in Urban and Environmental Planning Research at a consulting company, he became an assistant professor of Osaka University. His research fields include contemporary philosophy and the history of the western philosophy in eighteenth to nineteenth century in Germany and France. He is also interested in the philosophical approach of artifacts including the urban theory and aesthetic uses of language.

Yukari Nagai is the dean and a professor of School of Knowledge Science, Japan Advanced Institute of Science and Technology. B.A. in Art (1986), M.A. in Art (1989), Ph.D. from Chiba University (2002), and Ph.D. in Computing Sciences from University of Technology, Sydney (2009). Her research interests include creativity, design, art, and cognitive process, especially concept generation. She is also the Chair of Design Creativity SIG for the Design Society (2007–present).

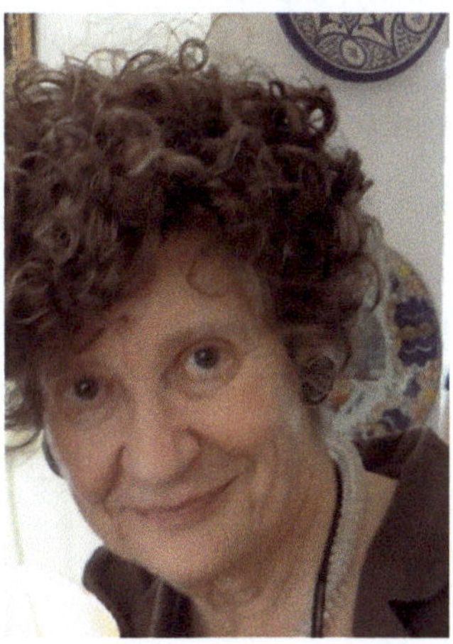

Professor Emeritus Gabriela Goldschmidt held the Mary Hill Swope Chair at the Technion—Israel Institute of Technology, Faculty of Architecture and Town Planning.

After graduating from Yale University she had practiced architecture until she became a full-fledged academic. She taught a large number of studio courses in architecture and industrial design and held visiting appointments at a number of universities. Her research encompasses design cognition, reasoning and representation, visual thinking and design education. She co-edited (with William Porter of MIT) *Design Representation*, published by Springer Verlag; her new book *Linkography: Unfolding the Design Process*, was published by MIT Press in 2014.

Hernan Casakin, is a Senior Lecturer in the School of Architecture, Ariel University, Israel. He holds a B.A. in Architecture and Town Planning from University of Mar del Plata, Argentina, and a M.Sc. and a D.Sc. in Architecture from the Technion, IIT, Israel. He had appointments as Research Fellow in the Department of Cognitive Sciences, Hamburg University, and the Environmental Simulation Laboratory, Tel Aviv University, and recently in the Faculty of Industrial Design Engineering, and the Faculty of Architecture, TUDelft, Delft University of Technology, Netherlands. He is a board member of several international journals. His research interest and publications are in design thinking and creativity.

Shulamith Kreitler was born in Tel-Aviv, has studied psychology, philosophy and psychopathology in Israel, Switzerland and the USA. She got her Ph.D. in Bern Switzerland. Has worked as a professor of psychology in Harvard, Princeton, Yale, and Vienna. She has been a professor of psychology at Tel-Aviv University since 1986. Has published about 200 papers and 10 books in motivation, cognition, psychopathology and health psychology. She has created the theory of meaning, and the cognitive orientation theory of behavior and wellness. Some of her publications are: Cognitive Orientation and Behavior (1976), The Cognitive Foundations of Personality Traits (1990), and Cognition and Motivation (2012).

Barbara Tversky is a cognitive psychologist who has worked on memory, categorization, spatial language and thinking, event perception and cognition, diagrammatic reasoning, creativity, and gesture. She has enjoyed working with linguists, computer scientists, geographers, philosophers, engineers, educators, biologists, chemists, artists, and designers on a range of projects. She is Professor of Psychology Emerita at

Stanford University and Professor of Psychology at Columbia Teachers College. She is a member of the American Academy of Arts and Sciences and a Fellow of the Cognitive Science Society, the Society for Experimental Psychology, and the Association for Psychological Science.

Shuichi Iwata is working for "Data and Society"-making data on science and technology available for everyone. President of CODATA (Committee on Data for Science and Technology/ICSU) (2002–2006), Professor at Graduate School of Project Design, Emeritus Professor of the University of Tokyo, Editor-in-Chief (Data Science Journal, 2008–), EAJ & SCJ Member. Born on January 29, 1948; 1975 Doctor of Engineering, Nuclear Engineering, the University of Tokyo; Award and Honors: Honda Memorial Young Researcher Award/Iketani Science Foundation Award/Promotion of Science and Technology Information Award, Japan Science and Technology Agency/Paper Award, The Japan Institute of Metals/ GIW Best Paper Award.

Amaresh Chakrabarti is professor at Centre for Product Design and Manufacturing, Indian Institute of Science (IISc), Bangalore. He holds B.E. in Mechanical-

Engineering (University of Calcutta), M.E. in Mechanical-Design (IISc), and Ph.D. in Engineering-Design (University of Cambridge UK). Before joining IISc, he led for 10-years the Design-Synthesis-team at the Engineering-Design-Centre, University of Cambridge. He has 10 books, over 250 articles, and 6 patents. He co-authored DRM, a design research methodology. He is on editorial boards of ten International-Journals. He was elected twice to Advisory-Board, Design-Society, and is currently on its Board-of-Management. He is chair for ICoRD-series, CIRP-Design-2012 and ICDC-2015, and Honorary-Fellow, Institution of Engineering Designers UK. Seven of his papers won top-paper-awards in international-conferences.

Noriko Hashimoto, born in 1948, graduated from the University of Tokyo (Faculty of Letters) in 1974. She received M.A. from after-graduated Course of Humanities of the University of Tokyo (1978). Assistant of the Faculty of Letters of the University of Tokyo (1981–1983). Secretary General of International Symposium of Eco-ethica (1981–2006), Secretary General of International Societies for Metaphysics (1999–2009), Secretary General of Tomonobu Imamichi Institute for Eco-ethica in Copenhagen (2007–). Vice-Director of Centre International pour l'Etude Comparee de Philosophie et d'Esthetique (2000–2012). She received Dr. of Religious Culture on Studies on Levinas in 2001. With Peter Kemp she published three volumes of "*Eco-ethica*". She was elected as CD member of FISP (Federation International des Societes de Philosophie) in 2012.

Russell Thomas is a Ph.D. student in the Department of Computational Social Science at George Mason University, and also Security Data Scientist at Zions Bancorporation. He has extensive experience in the computer industry, including new product development, manufacturing, marketing, IT, and business process improvement. His research interests include innovation and risk management in socio-technical systems, cyber security, agent-based modeling, and machine learning to support business intelligence. He is a frequent presenter at industry and academic conferences, and is a founding member of the Society of Information Risk Analysts (SIRA).

John Gero is a Research Professor at the Krasnow Institute for Advanced Study and at the Department of Computational Social Science, George Mason University and in Computer Science and Architecture at the University of North Carolina, Charlotte. Formerly he was Professor of Design Science and Co-Director of the Key Centre of Design Computing and Cognition, at the University of Sydney. He is the author or editor of 50 books and over 600 papers and book chapters. He has been a Visiting Professor at MIT, UC-Berkeley, UCLA, Columbia and CMU in the USA, at Strathclyde and Loughborough in the UK, at INSA-Lyon and Provence in France and at EPFL-Lausanne in Switzerland.

Gaetano Cascini, holds a Ph.D. in machine design and is Associate Professor at Politecnico di Milano since 2008. His research interests cover Design Methods and Tools with a focus on the concept generation stages both for product and process innovation. He is member of the Editorial Board of the Journal of Integrated Design & Process Science and of the International Journal of Design Creativity and Innovation. He is also member of the Advisory Board of the Design Society and co-chair of its Design Creativity SIG. He has coordinated several research projects and currently is the coordinator of the Marie Curie-IAPP FORMAT project. He has authored more than 100 papers presented at International Conferences and published in authoritative Journals and 12 patents.

Francesca Montagna, Ph.D, is assistant professor at Politecnico di Torino. Research topics range from Management of Innovation, Engineering Design and Decision Making. Lecturing is mainly carried out in the curriculum degree of Industrial Engineering where she teaches "Innovation management and product development". She often teaches in diverse master degree courses for public and private institutions, such as LUISS in Rome. She has authored or co-authored diverse academic papers and usually serves as reviewer for academic journals. Often she is expert evaluator of funding project proposals at national and international level. She is member of A.I.TE.M., Design Society (http://www.designsociety.org). She usually serves as reviewer on conferences of the Design Society (e.g. ICED, DESIGN conferences) and IFAC (e.g. INCOM) or CIRP.

Pascal Le Masson is Professor at MINES ParisTech, Chair of Design Theory and Methods for Innovation. He is the Director of the Center for Management Science. His research unfolds in three main directions: i/design theory (C-K theory, mathematical foundations,...) ii/collective innovative design methods (creativity, prototyping, user involvement processes) iii./innovative design organization (advanced R&D), iv/economics of design. He has published several papers and a book "Strategic Management of Innovation and Design" (co-authored by Armand Hatchuel and Benoit Weil, Cambridge University Press). He co-chairs (with Eswaran Subrahmanian and Yoram Reich) the "Design Theory" Special Interest group of the Design Society.

Benoit Weil is Professor at MINES ParisTech, Chair of Design Theory and Methods for Innovation. His research areas are (1) Design Theory and models for Design

Science; (2) Design and R&D Management; (3) Management of the Innovative Firm; (4) Theory of Design Regimes; (5) R&D and Design history. He has published several papers and a book "Strategic Management of Innovation and Design" (co-authored by Armand Hatchuel and Pascal Le Masson), Cambridge University Press).

Olga Kokshagina is Ph.D. candidate at Center for Management Science, Mines ParisTech School, France. Her research focuses on the management of innovative design capabilities and more particularly on the areas of innovation and technology management, projects portfolio management, uncertainty management and R&D tools and methods. Her thesis dissertation models the design of generic technologies and investigates the associated models of collective action. Being interested in the specifics of innovative processes in high-tech industries such as nano-, biotechnologies, semiconductors, IT, telecommunication, she has conducted her dissertation work in collaboration with a leading European semiconductor company—STMicroelectronics.

Rivka Oxman is an Associate Professor at the Faculty of Architecture and Town Planning where she has been a Vice Dean. She holds D.Sc. degree from the

Technion IIT. Her research contributed to cognitive and computational modeling of design and currently focusing on the exploration of emerging digital technologies and their impact on design. She has been a Visiting Professor at Delft University of Technology; Stanford University; and Visiting Scholar at MIT and Harvard University. She was appointed as Fellow of DRS for her contributions to design research. Her recent publications are: The New Structuralism: Design, Engineering and Architectural Technologies; John Wiley (2010); and Theories of the Digital in Architecture, Routledge (2014).

Contributors

Hernan Casakin Ariel University, Ariel, Israel

Gaetano Cascini Politecnico di Milano, Milan, Italy

Amaresh Chakrabarti Indian Institute of Science, Bangalore, India

John Gero George Mason University, Fairfax, VA, USA and University of North Carolina at Charlotte, Charlotte, NC, USA

Gabriela Goldschmidt Technion—Israel Institute of Technology, Haifa, Israel

Noriko Hashimoto Aoyamagakuin Women's Junior College, Tokyo, Japan

Michio Ito Tokyo University of Agriculture and Technology, Tokyo, Japan

Shuichi Iwata Graduate School of Project Design, Tokyo, Japan

Olga Kokshagina MINES ParisTech, Paris, France

Shulamith Kreitler Tel Aviv University, Ramat Aviv, Tel Aviv, Israel

Udo Lindemann Technical University Munich, Garching, Germany

Pascal Le Masson MINES ParisTech, Paris, France

Francesca Montagna Politecnico di Torino, Torino, Italy

Yukari Nagai Japan Advanced Institute of Science and Technology, Ishikawa, Japan

Rivka Oxman Technion—Israel Institute of Technology, Haifa, Israel

Toshiharu Taura Kobe University, Kobe, Japan

Russell Thomas George Mason University, Fairfax, VA, USA

Barbara Tversky Stanford University, Stanford, CA, USA and Columbia Teachers College, New York, NY, USA

Benoit Weil MINES ParisTech, Paris, France

[illegible], Kobe, Japan

Part I
General and Philosophical Consideration on Pre-design, Design, and Post-design

Chapter 1
The Design of Technology: Bridging Highly Advanced Science and Technology with Society Through the Creation of Products

Toshiharu Taura

Abstract This chapter describes the concept of the Design of Technology. The role of the Design of Technology is to create a bridge between *highly advanced science and technology* and society through the creation of products. First, this chapter describes the differences that exist between two types of design: (1) design that employs *conventional science and technology*, and (2) design that employs *highly advanced science and technology*. Second, it describes the processes in which science and technology and its products become accepted by society. In addition to the conventional methods specialists use to provide simple explanations of physical principles, the chapter highlights three additional views: (1) science and technology's infiltration of the culture of society, (2) the organization of the social system to cope with negative aspects of science and technology, and (3) the discovery of new mechanisms to recognize unimaginable and invisible dangers. Based on the above discussion and analysis, the chapter discusses methods that can be used to approach the Design of Technology. It suggests that designers and researchers should adopt more open and accepting attitudes and focus on meaning rather than procedure.

Keywords Culture • Design • Society • Technology

1.1 Introduction

Our living environment is filled with products that include science and technology. It would not be an exaggeration to state, "We cannot discuss "design" without discussing science and technology." Products that include science and technology have provided significant benefits to society. On the other hand, these products have simultaneously caused significant problems. In this chapter, the term "science and

T. Taura (✉)
Kobe University, Kobe, Japan
e-mail: taura@kobe-u.ac.jp

© Springer Japan 2015
T. Taura (ed.), *Principia Designae – Pre-Design, Design, and Post-Design*,
DOI 10.1007/978-4-431-54403-6_1

technology" refers to disciplines that rely on fundamental knowledge of natural phenomena to realize product requirements or specifications. The term, "products," is used to describe equipment or plants (e.g., power stations or buildings), as well as other items used by consumers (e.g., computers or cars). The term, "society," is used to describe the individuals who use those products.

Let us examine the concept, "a car." Today, "a car" is a globally used product. Many cars are designed and produced on a daily basis. However, many individuals have been killed in car accidents. Why, then, do we use such a dangerous artifact? We have not yet explicitly discussed or determined the degree to which dangers related to the use of a car might be considered acceptable.

On the other hand, in March 2011, Japan experienced "a nuclear power station" disaster. The nuclear power station had already been designed and produced. However, the fundamental reasons why we create and use products that present significant dangers have not yet been clarified. Thus, social feelings related to the acceptance of these types of dangerous stations have not yet been developed.

What differences exist between "a car" and "a nuclear power station?"

We can discuss this issue by focusing on the gap that exists between *conventional science and technology* and *highly advanced science and technology*. In this context, the term, *conventional science and technology*, refers to traditional science and technology applied to the production of products (e.g., dynamics, fluid dynamics, thermodynamics, electrical engineering, chemical engineering, and so on). *Highly advanced science and technology* is a term applied to science and technology that has advanced over the past two or three decades (e.g., nuclear science, genetic engineering, nanotechnology, ICT, and so on). The author will describe the differences in scope covered by each design: (1) design that relates to *conventional science and technology*, and (2) design that relates to *highly advanced science and technology*. The concept of "a car" is employed to exemplify design that relates to *conventional science and technology*. Here, focus is placed on the car's fundamental technology, even though some *highly advanced science and technologies* have been implemented in "a car." In this chapter, the author will demonstrate that design in the near future must more accurately relate to a wider scope that encompasses both *highly advanced science and technology* and society. Further, the author will suggest that designers and researchers engaged in *highly advanced science and technology* should adopt more open and accepting attitudes to related fields, such as cultures, social systems, and so on. They must also consider all aspects from the viewpoint of "meaning" rather than "procedure."

1.2 Pre-design, Design, and Post-design

The conventional design process is comprised of the following stages: conceptual design, embodiment design, and detailed design. The author believes a simple examination of these three design stages cannot capture the essential nature of the relationship between science and technology and society; this is because the

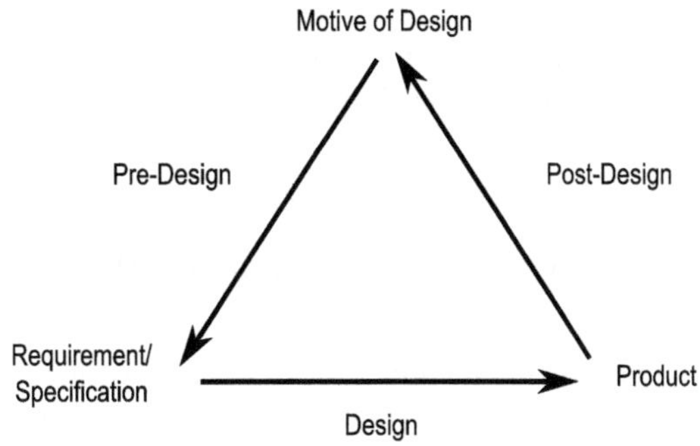

Fig. 1.1 The pre-design phase, the design phase, and the post-design phase

relationship is generated outside of the three stages. Thus, it is important to examine the design phases that occur prior to and after these three stages occur. The author refers to the former phase as the Pre-Design phase. He refers to the latter phase as the Post-Design phase (Taura 2014). Further, the author proposes the conception of *motive of design* that implies "an underlying reason for the design of a product" (Taura 2014). Based on consumer-product interactions (experiences of utility or accidents), or, in some cases, without reliance on consumer-product interactions, consumers' feelings and criteria for science and technology and its products, or their awareness of problems with existing science and technology and its products, are overtly or covertly created and contained in each person or in society; *personal motive* is that which exists deep in a person's mind and *social motive* is that which is shared in society. These newly formed feelings, criteria, or this awareness of problems constitute the *motive of design*. They become the reasons to design new products. The *motive of design* is different from so-called needs. "Design" occurs in a situation in which needs do not yet exist. The *motive of design* exists underneath the "needs" and generates them.

The *Design phase* is the so-called conventional design phase. It is comprised of three stages: conceptual design, embodiment design, and detailed design. During this phase, a new product is developed that will attempt to satisfy the requirements or specifications explicitly proposed for new products. This phase is described as a process of "implementation" during which science and technology is implemented into products to meet requirements and specifications.

The *Post-Design phase* is a process in which the *motive of design* is created and stored in an individual or in society through consumer-product interactions, or, in some cases, without reliance on consumer-product interactions.

The *Pre-Design phase* is a process in which the concrete requirements or specifications for new products that society might accept are created from the *motive of design*.

Figure 1.1 illustrates the relationship that exists among the Pre-Design phase, the Design phase, the Post-Design phase, and the *motive of design*. As noted in Sect. 1.1, this chapter focuses on the role of science and technology implemented in products. From the viewpoint of science and technology, the *motive of design* may be driven by developments in science and technology. In addition, the *motive of design* may be created during the Post-design phase when the benefits (positive aspects), and dangers or anxieties (negative aspects) caused by science and technology are recognized by society when they use the products.

1.3 Science and Technology, Products, and Society

To discuss the nature of design from the viewpoint of science and technology, we must determine the relationships that exist among science and technology, products, and society.

First, the relationship that exists between products and society involves a process in which products provide society with "service" and the products receive "interpretations" (evaluation and requirements) provided by society. In this case, "service" involves the mental contributions made to society through individuals' expansion and recall of feelings (e.g., musical instruments), as well as physical contributions made to society by increases in efficiency and safety (e.g., a car).

Second, the relationship between science and technology and products involves a process in which products are provided with functions by science and technology, and the products then return evaluations and requirements to science and technology. In some cases, this kind of science and technology might be described as "technology with a narrow view." Alternatively, in many cases, the entirety of science and technology and its products can be described as "technology." Furthermore, the entirety of science and technology, products, and society can be described as "technology with a wide view." Figure 1.2 provides an illustration of the relationships that exist among science and technology, products, and society. This figure indicates that the role of products is to apply knowledge provided by science and technology to society. The role of design is to create a bridge between science and technology and society through the creation of products.

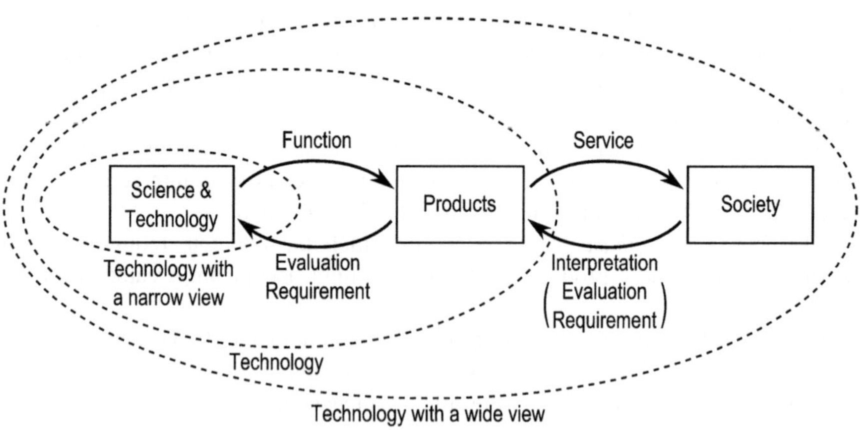

Fig. 1.2 Science and technology, products, and society

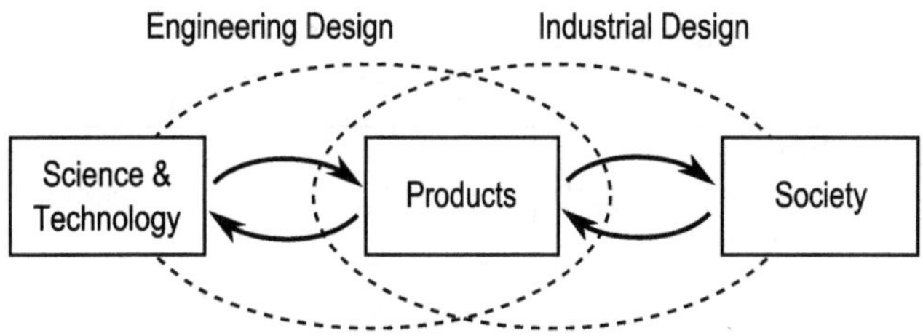

Fig. 1.3 Engineering design vs. industrial design

Based on the framework illustrated in Fig. 1.2, it is possible to distinguish the differences that exist between so-called industrial design and so-called engineering design (see Fig. 1.3). So-called engineering design is located closer to science and technology. So-called industrial design is located closer to society.

It is also possible to distinguish between design that relates to *conventional science and technology* and design that relates to *highly advanced science and technology* by applying the above-mentioned framework. As illustrated in Fig. 1.4, when *conventional science and technology* is implemented in products, *conventional science and technology* and its products are located close to one another. In other words, the product designer and the *conventional science and technology* researcher are employed by the same person or persons who belong to the related organization.

For example, "a steam locomotive" or "a car" was designed and produced by the same company that conducts research on the fundamental technology required to develop a new product. In other words, both science and technology and its products are included in the scope of "design." This "design" is referred to as *conventional engineering design*. The role of *conventional engineering design* is to realize product requirements or specification through developments in science and technology. In these types of cases, science and technology is understood (interpreted) by society through product-consumer interactions (see Fig. 1.4).

On the other hand, in cases in which *highly advanced science and technology* is implemented in products, both *highly-advanced science and technology* and its products remain separate from one another because additional specialization progresses in accordance with additional advances that occur in science and technology. For example, "a nuclear power station" is designed and produced by a company that differs from a company that conducts research on fundamental technology required for the development of a nuclear power station. In other words, only the product implementation process is included in the scope of the "design." This design is referred to as *modern engineering design*. The role of *modern engineering design* is to implement *highly advanced science and technology* into products to realize their requirements or specifications. In some cases, *highly advanced science and technology* can directly induce anxiety in society. This often occurs because individuals find these products unimaginable and invisible. They might also struggle when they imagine the possible magnitude and longitude of the damage these products might cause. For example, nanotechnology, nuclear science, or genetic engineering can induce anxiety in society because they are invisible. In addition, they can cause damage of great magnitude and longitude (see Fig. 1.5).

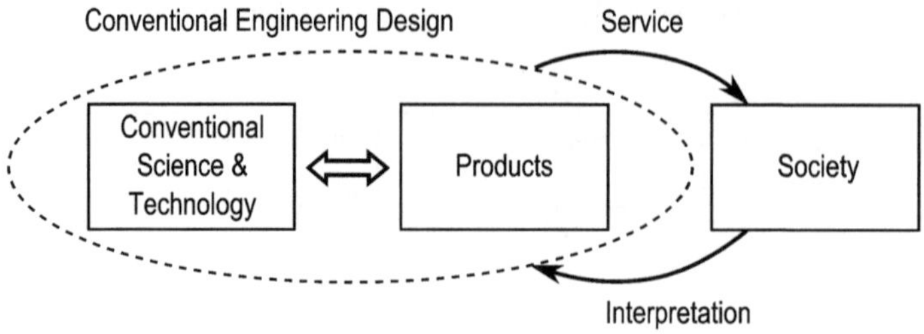

Fig. 1.4 Conventional engineering design

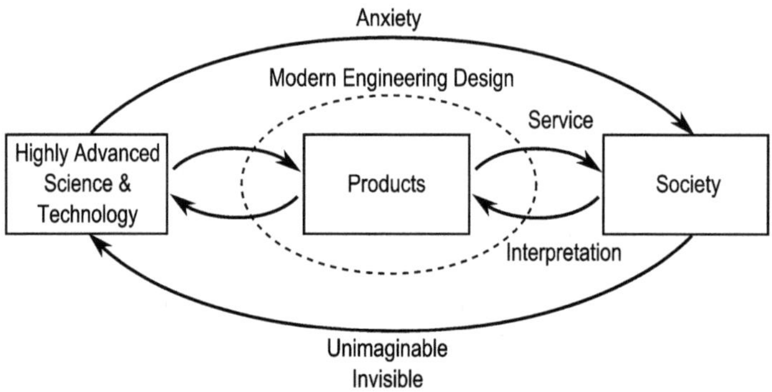

Fig. 1.5 Modern engineering design

1.4 In What Ways Are Science and Technology and Its Products Accepted by Society?

It is important to note that both science and technology and its products involve negative, as well as positive aspects. In other words, although science and technology and its products might provide efficiency and comfort to society, they might also cause damage to individuals. For society to accept science and technology and its products, society is thought to accept both their negative and positive aspects in a fundamental way as well as through considering its superficial merit or lack of merit.

One conventional process by which science and technology and its products can be accepted by society is through specialists' efforts to explain their mechanisms by the use of plain language or the use of statistics. For example, the physical principles of thermodynamics, genetic manipulation, or nuclear fission can be explained by the use of illustrations. The risks involved in these technologies can be explained by the use of probability. Although these processes are certainly useful, the author believes additional views may exist that might affect whether science and technology and its products will be accepted by society.

First, it should be noted that science and technology and its products become often involved in cultures within society. Certainly, many descriptions of science and technology and its products have appeared in art and literature. These descriptions help expand the limited and expected functional aspects usually recognized by society. For example, Claude Monet, the Impressionist artist, painted a representation of a steam locomotive. This fact that this artist chose to paint this subject indicates that the steam locomotive was naturally accepted on a deep feeling level by society. A steam locomotive is a typical product that involves *conventional engineering design*. Further, the author suggests the Japanese sword as an exemplar of a product that is a work of art, as well as an outcome of *conventional science and technology*. A Japanese sword is a curve-bladed weapon that has a single cutting edge. The creation of a Japanese sword is a specialized traditional process that involves heating and hammer-forged hardening. This type of sword is characterized by its great spring-like toughness and its extremely sharp cutting edge. Recently, Japanese swords are primarily considered art objects. They are often treasured for the sense of tranquility apparent in their cold, serene beauty. In cases in which products are designed through *conventional engineering design*, science and technology might be rather acceptable to the culture of a society because science and technology and its products are located close to one another and are both visible to the society. Further, science and technology is believed to act as a leader of the culture of that society. In fact, science and technology causes individuals to expand and recall feelings. This ultimately stimulates and enhances the culture of the society. For example, bronze casting technology expanded artists' abilities for self-expression. It enabled them to create impressive art that could inspire deep feelings.

Alternatively, in cases in which products are designed through *modern engineering design*, it might be difficult for *highly advanced science and technology* to gain acceptance from the culture of a society because it remains separate from products and it remains unimaginable and invisible to society. It can be difficult to paint or describe *highly advanced science and technology*. However, it is not necessarily impossible to gain acceptance from society, even if they cannot be painted or described. For example, the physical principles of many electrical products, such as televisions, personal computers, lighting equipment, and so on, are invisible. However, they are used naturally in society. In addition, some electrical products, such as personal computers, cell phones, and lighting, are believed to have the potential to expand human minds. Some believe these products can be interpreted by society because they appeal to individuals and cause them to recall deep feelings. For example, many products developed by Apple Inc. appear to expand the analogue world by the use of digital technology. However, it may take some time before *highly advanced science and technology* might become naturally accepted on deep feeling levels by society.

Second, it is important that a particular social system be developed to cope with the negative aspects of science and technology and its products. As noted above, "a car" is a product used globally. Many cars are designed and produced on a daily basis. However, many people have been killed in car accidents. Why do we use these dangerous products? The author believes this can be explained by two implicit manners. The first manner relates to the notion of responsibility. The author believes that the responsibility for car-related dangers can be completely divided into smaller responsibilities. Each responsibility can then be adopted by an individual or a company. For example, the responsibility for cars can be divided into the following areas of responsibility: design, manufacturing, maintenance, driving, and so on. Further, each responsibility can be divided into smaller, more-detailed responsibilities. It is important to note that the term, "responsibility," came into use in society after the Industrial Revolution began. The second manner relates to the notion of risk. When we apply the notion of "risk," we can observe feelings of "danger" in an objective manner. When we observe feelings objectively, rather than subjectively, we can address the negative aspects of cars rationally (e.g., car insurance). In cases in which products are designed through *conventional engineering design*, science and technology might be rather more acceptable to society because of the organization of the social system that involves the concepts of responsibility and risk. However, in cases in which products are designed through *highly advanced science and technology*, the social system that corresponds to responsibility and risk in *conventional science and technology* has not yet been organized. For example, with respect to nuclear power stations, the current social system that corresponds to responsibility and risk has not yet discovered ways to cope with related problems. In many cases, problems caused by *highly advanced science and technology* cannot be captured within the current framework of society. In some cases, these problems should be captured in a framework that extends beyond the current age. A characteristic of *highly advanced science and technology* is that it can cause problems that will continue over a long period. Who will assume responsibility for future problems? The people who should assume responsibility will no longer exist when these problems arise. Furthermore, how responsibility and risk is actualized? Normally, responsibility and risk are methods used to transfer abstract or subjective feelings related to the negative aspects of engagement so those feelings can be accepted in objective ways. These transformations become actualized through feedback obtained from experiences of actual accidents or failures. However, in *highly advanced science and technology*, accidents or failures are unacceptable because they may cause significant damage that extends beyond society's ability to recover. We must establish new notions and methods that will extend current conceptions of responsibility and risk to a degree sufficient to accept dangers related to *highly advanced science and technology* without relying on feedback obtained from experiences of actual accidents or failures.

Third, it should be noted that mechanisms to recognize science and technology and its products play an important role for society to accept them. *Conventional science and technology* is often perceived visually. For example, we can perceive the movement of a steam locomotive. Although we cannot see inside the steam boiler, we sense the reality of the principle of steam locomotion by viewing the movement of the wheels and the smoke. However, some types of *highly advanced science and technology* are unimaginable and invisible. For example, radioactive rays are both dangerous and invisible. Furthermore, we must imagine the 10,000- or 100,000-year period during which nuclear reactor waste should remain in storage. The period of time we are asked to imagine extends beyond the capacity of our imagination.

However, we can recognize unimaginable and invisible things indirectly by using abstract notions. For example, we calculate "force" when we design the mechanism of a machine. However, the force of classical mechanics cannot be directly observed. It is calculated from the displacement or acceleration of an object. It can be said that force is an abstract notion created by humans that is beneficial for understanding physical phenomena. In addition, "the future" is an abstract notion created by humans through language. We can never draw an exact picture of the future. We can imagine what things might be like in the future, but it is impossible to visualize a precise notion of the future itself. We think this kind of highly abstract notion can be represented in language. The above-mentioned discussion suggests a possibility to recognize the notion of a 10,000- or 100,000-year period by using language.

Yet, some invisible dangers of *highly advanced science and technology* are used naturally by society. For example, Induction Heat (IH) cookers or microwave ovens are used daily even though these products may cause accidents if they are used incorrectly. These products are not solely considered acceptable by society because of their recognitive interpretation or their cultural acceptance. Rather, they became acceptable because of the amount of experiential usage that accumulated after rational development based on scientific and technological knowledge was performed to ensure their safety.

1.5 The Design of Technology

Based on the discussions presented above, the author proposes the concept of the *Design of Technology*. The Design of Technology extends its view on the object of design from the products themselves to include the scope that encompasses both *highly advanced science and technology* and society. Thus, the role of design is to create a bridge that connects *highly advanced science and technology* and society through the creation of products. Further, the Design of Technology infiltrates the culture of society and it requires the organization of a social system that can cope with the negative aspects of science and technology and the new mechanisms to recognize unimaginable and invisible things indirectly (see Fig. 1.6). Recently, designers have limited their perspectives during the product actualization process by relying primarily on their knowledge of systems engineering or plastic methods. To realize the Design of Technology, designers or scientists engaged in product design or research related to *highly advanced science and technology* must expand their views. Here, the author does not insist that these professionals become artists or sociologists. Rather, the author believes it is important to focus on "meaning" rather than "procedure." In an engineering design school, students are trained to learn the procedures to calculate stress, displacement, calorie, flux, and so forth.

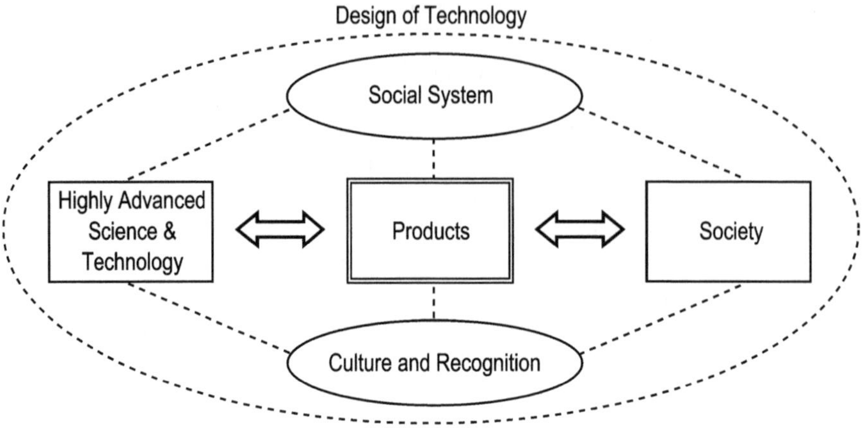

Fig. 1.6 The design of technology

Students are evaluated according to whether they can perform the procedures correctly, and the meaning of the procedures is not discussed.

How might individuals learn to focus on meaning rather than procedure?

First, individuals must understand that so-called natural laws are nothing more than knowledge used to explain phenomena. These laws differ completely from the truth. We normally pursue simpler explanations of physical phenomena. At times, we believe the simplest explanation is true. For example, although the primary difference between the Ptolemaic and Copernican systems is the degree of simplicity and richness of explanation involved, we sometimes believe the Copernican system is true and the Ptolemaic system is false. However, it is "true" that we see the "sunset" and "sunrise." We do not say "earthrise" or "earthset." If these phenomena are described as "earthrise" or "earthset," the phenomena might be difficult to paint or set to music. It is important to note that knowledge obtained by science and technology based on actual phenomena is only effective when that knowledge is used to explain particular phenomena. Much less, this knowledge is not sufficient to control them. In other words, the products created based on this knowledge should manifest additional behaviors that exceed designers' expectations or intentions. In reality, after many experiences with accidents or problems, these products may begin to manifest required stable behaviors under limited and predefined conditions. When these conditions change or when products are revised even under the same conditions, the products might necessarily manifest unexpected behaviors, even if those behaviors were precisely examined in advance based on available knowledge. This is not a defect of science and technology. Rather, it is its essential nature.

Second, designers should investigate whether the preconditions that affect reasoning are acceptable. In many cases, designers realize requirements or specifications in products by implementing science and technology. A number of preconditions are involved in these processes. However, these preconditions are usually obtained from outside sources. Designers often do not believe they must examine the validity of these preconditions. This consideration can be illustrated by the use of an expression to describe reasoning "A → B." Here, according to formal logic, in a case in which "A" is false, "A → B" is defined as true. The above-noted definition implies if the precondition is false, then the reasoning process itself is always true. It appears that this implication involves views similar to the attitudes of the designers discussed above. These conservative attitudes will not be inviting to professionals in related fields. In addition, they will not assist in making products and the *highly advanced science and technology* implemented in those products more easily understandable to society.

Third, designers should consider placing the *motive of design* within the wider scope of the Design of Technology, rather than in the conventional narrow scope of products. However, this does not imply that we must focus on superficial problems located right before our eyes. Rather, we should focus on essential issues that operate on levels below our immediate attention. The author believes that each individual's "inner motive" is related to his/her willingness to engage in deep consideration of these factors. Here, the "inner motive" is a drive that exists in each designer's mind. This drive is related to feelings and criteria for products. Alternatively, the "outer motive" is a drive contained in products. That drive can be related to each designer's awareness of problems (Taura 2014). The author believes that the "inner motive" can also be related to each designer's willingness to consider preconditions, or to his/her willingness to adopt an open and accepting attitude towards related fields, as noted above.

1.6 Discussion

Is the Design of Technology possible? To achieve the Design of Technology, the following issues must be addressed.

First, how *highly advanced science and technology* become naturally accepted at deep feeling levels in the culture of society; and/or

Second, how new concepts and methods are developed to extend the current conceptions of responsibility and risk so that society can adequately cope with the anxieties induced and dangers inherent in *highly dvanced science and technology*; and/or

Third, how new methods to recognize and manage the magnitude and longitude of unimaginable dangers associated with *highly advanced science and technology* are discovered.

The author wonders whether these three issues were addressed repeatedly during processes employed in the development of earlier technologies. If this is the case, then time will solve these problems. A car might not have been acceptable to society when it first appeared. Yet, it may have become acceptable after many consumer-product interactions occurred (experiences with its utility or experiences with accidents). We might assume that the social system that copes with car accidents was not organized when the car was first introduced.

Yet, the author believes that the Design of Technology involves issues that differ from the issues that affected previous cases.

First, a number of *highly advanced science and technologies* involve completely unimaginable notions (i.e., invisible dangers of significant magnitude and longitude).

Second, accidents and failures are unacceptable for a number of *highly advanced science and technologies* because they might cause such significant damage that society would be unable to recover.

These two issues are extremely difficult to approach. To address the first issue, we must wait for society to accept these notions. This will take a long time because society must create and share completely new views. The second issue appears impossible to address. As mentioned in the previous section, we cannot control the behaviors of products. Stable and safe behaviors developed after many experiences with accidents or problems occurred. The author believes it is essentially impossible to obtain product stability and safety without the occurrence of experiences with accidents or problems. Further, the social system that copes with negative aspects of science and technology has been organized based on experiential accumulation. It seems almost impossible to organize this type of system in advance without relying on feedback obtained from actual experiences.

In this chapter, comparisons were made between "a car" and "a nuclear power station." It is important to note that these concepts cannot be aligned on the same level. Rather, "a nuclear power station" should be compared with "a thermal power station." Even when we make comparisons in this case, the concept, "thermal power station," is similar to the concept, "steam locomotive." However, if we focus on *all* cars or thermal power stations used on the earth, we know they emit a great deal of carbon dioxide, which is believed to cause global warming. Global warming could cause significant damage from which society might be unable to recover. This discussion suggests that even *conventional science and technology* will display the same characteristics as *highly advanced science and technology* when its products are used in large quantities.

Currently, if we engage in the Design of Technology, we should remember that some essential issues remain unsolved. We must hope that intellectuals from many fields (e.g., designers, researchers, artists, novelists, sociologists, philosophers, etc.) will collaborate to develop and orchestrate new harmonious knowledge.

Acknowledgments The conceptions introduced in this chapter were based on discussions I had with Prof. Michio Ito (Tokyo University of Agriculture and Technology, Japan) and Prof. Yukari Nagai (Japan Advanced Institute of Science and Technology). The author would like to express deep gratitude to them both. The idea that the term, "responsibility," came into use after the Industrial Revolution began was based on a comment made by Tomonobu Imamichi during a personal conversation we had in 1996. The concept of the differences that exist between risk and danger developed from a comment made by Michio Ito during a conversation we had in 2012. His comment was based on the work of Ulrich Beck.

Reference

Taura T (2014) Motive of design: roles of pre- and post-design in highly advanced products. In: Chakrabarti A, Blessing L (eds) An anthology of theories and models of design. Springer, London, pp 83–98

References

Chapter 2
Supporting Creativity in Engineering Design: A Position Paper

Udo Lindemann

Abstract Within this paper it is not the aim to explain the mechanisms of creativity, possibilities to measure creativity or to deliver a definition.

Based on some experience based on personal involvement and a large number of case studies some conclusions will be presented. There are also a few critical remarks concerning creativity related practice as well as research.

Creativity is a very complex topic! A differentiation between different kinds of creativity is required as well as an understanding of the dependencies between problem, situation, acting people as a minimum. Based on that a preliminary model may be established, which should serve formulating hypotheses for further empirical research. The outcome of this research may help to improve the model.

Creativity is not just inspiration; it is hard work and should be supported by systematic procedures and methods.

Keywords Creativity • Creativity and thinking • Evaluation of creativity • Systematic procedures

2.1 Design and Pre-design and Post-design?

Engineering Design has to generate the description/documentation of solutions/ products for the society. This has to be done in close collaboration with other departments and roles within the processes of generation these solutions/products. Where is the starting point and where/when does it end?

Engineering design also has to improve given products by eliminating weaknesses, raise the efficiency, reduce cost, etc.

The phase before specification/requirements is quite often dominated by marketing sales and other departments. But engineering design has to be part of it, which was shown and argued in a number of research projects as well as in industrial practise. So there should not be any real pre-design-phase, as long as we discuss design (product development) cycles in a holistic way and not as a

U. Lindemann (✉)
Technical University Munich, 85748 Garching, Germany
e-mail: udo.lindemann@tum.de

© Springer Japan 2015

T. Taura (ed.), *Principia Designae – Pre-Design, Design, and Post-Design*,
DOI 10.1007/978-4-431-54403-6_2

specific project. And similar is the point of post-design. As long as products are on the market and used by customers engineering design at least has the obligation to observe the products in the sense of preventing risk.

Post-design may also be understood as the beginning of Pre-design of the next product generation including improvements, new and additional features, addressing additional market segments and so forth.

Engineering design has to play an active role in the whole range of phases from society to society!

Engineering design has to provide (new) solutions/products to the markets and to the society. Based on these points engineering design is fulfilling its role within the supplier-network to generate sales and profit.

Creativity is discussed as one of the important drivers of improving solutions/ products and generating new solutions/products. But creativity is also required to set the "right" goals and requirements, to find potentials for cost reduction, to solve problems regarding reliability, to organize engineering processes etc. It seems that there have to be different categories of creativity.

2.2 Creativity

There is a lot of literature dealing with creativity; but do we really understand what creativity is about?

2.2.1 Support Creativity

Higgins (1994) in his book about Creative Problem Solving stated that for him Brainstorming is one of the most successful methods fostering creativity. Other authors like Furnham (2000) come to a completely different result.

The author discussed this matter with a number of consultants working in the field of creative problem solving. Most of them stated that brainstorming should be seen as one of the weakest creativity supporting methods. Talking with practitioners and observing them during daily business brainstorming seems to be pervasive in industry.

Confronted with this situation the author discussed the contradiction of often used vs. lack of efficiency with a senior colleague in psychology. At the end of this discussion we agreed on a list of numerous steps and actions to be processed for a successful and efficient process of supporting creativity. But still there are doubts that group processes are really helpful in generating ideas.

After discussing only one of a large number of critical examples for supporting creativity it seems absolutely necessary to develop a much better understanding of creativity!

2.2.2 Measure and Evaluate Creativity

There are a number of trials to measure creativity. Usually the attempts are focussing on the output of a creative process.

Let us discuss the question of measuring or evaluating creativity. The mostly used terms are *novelty* and *usefulness*. Looking at an example out of history we can find an important innovation for safety measures in cars, the air-bag system. The origin of these systems was an idea that leads to a patent in 1951. At that time this idea was really novel, but not useful at all. Why not? There was no adequate sensor, no adequate gas generator, no adequate airbag, and you may proceed with these points. It was just a nice invention at that time. Thirty year after this patent was published the first air-bag systems were available for customers. So the air-bag-system really became an innovation about 40 years after its invention.

The discussion of novelty and usefulness leads to some general questions: Novelty for whom and when? Usefulness was given for whom and when? These measures are at least very subjective and depending on time and situation!

Just two further aspects may be discussed:

The *number of ideas*: Why should a large number of ideas be an indicator of outstanding creativity? One brilliant idea may be enough!

The *grade of elaboration*: An excellent idea may be quite fuzzy at the beginning and elaborated by other people later on!

A story out of history will underline these two aspects: a control unit for a new automatic transmission had to be developed. State of the art at that time was a hydraulic system, which included valves and elements like impedances defining the dynamic behaviour. The main part of these controls was a casted and machined part. Because of time pressure and the knowledge about difficulties regarding late changes of these control units a creative engineer suggested to develop and to build an electric device. His argument at that time was just simple: exchanging a capacitor is much easier than changing the profile of a hydraulic channel. Conclusion: only one idea and not elaborated at all! The outcome was the first electronic control unit of an automatic transmission in the automotive industry.

In which kind of a situation is creativity required? Let us discuss three different situations as examples. Within the *first* situation we have to generate some requirements for a future solution that will be marketed after 3 years. What are the targets to be set based on fuzzy pictures of the future in markets, available technology or legislation? Within the *second* situation we have to develop an advanced functionality for the future version of our given product. Our company expects solutions to be protected by a set of patents. And our *third* situation happens to take place only 2 weeks before an important exhibition, where we are going to present our new product. Unfortunately one component failed during testing. Now we have to find a solution under high time pressure.

These examples just shall give hints that there is not only one kind of creativity, but that the specific kind of creativity required is depending on the specific situation. Measurement of creativity should be compared with measurement of thinking in general!

2.2.3 *Mechanisms of Creativity*

A more important question is, if the points of evaluation help us in understanding the mechanisms of creativity. There are a lot of definitions; there is a lot of literature with origin in scientific research, consultancy and others.

There are a number of different theories about creativity available, which try to explain at least specific details or aspects of creativity. But still we have no proved way of teaching and training of creativity—and we are not sure that these measures are possible at all.

A critical voice, W. Erharter, came up in 2012 with a book claiming that there is no creativity at all (Erharter 2012). In his book the author addresses different kinds of creativity like an individual characteristic, a capability, and a kind of behaviour, a problem solving technique, a kind of childish fantasy, a basis for innovation and others. Erharter seems to be a little bit provoking, but at the end he is talking about generating/creating solutions based on basic mental processes.

It seems that there are some basic mechanisms to support creating new things, we may call it creativity. Following the discussion up to now it seems that there may be different kinds of creativity and different ways of supporting these.

Based on experience we can state that there are proved ways of reducing and avoiding creativity by education or culture. Preventing children from doing some creative things or even punishing them if they start to be creative will have long term negative impact. Establishing a company-culture attacking individual motivation, punishing all failures, working without a clear set of rules way etc. will eliminate any creative spirit.

2.3 Engineering Design

In engineering design processes we have to create results, solutions, products. And there are usually a large number of targets, requirements, constraints etc. we have to keep in mind. This may be compared with navigation in a jungle with only limited infrastructure and lack of overview and insights.

First of all we have to identify the area of problems/demands we should work on. In this phase we have to cooperate with colleagues from strategy, product management, marketing etc. Having done this we have to elaborate the requirements, functions, constraints etc. of the future solutions/products. Here we have to have the "right" ideas for a successful realisation in future.

Based on these points the next phase may be planned and budgeted and the concretisation can go on. In this phase we have to collaborate intensively with production, purchasing and sub-suppliers, controlling etc. We have to have good ideas to come to successful solutions, to handle all the contradictions and the constraints.

After defining and documenting the solution/product engineering design has to optimise and improve or upgrade the solution/product. In addition there is an obligation to observe the product in the market from the point of risk management. Nevertheless the kind of use, impact and acceptance in the market/society is an important basis for the next generation of development.

2.4 Support Engineering Design Processes

In general Graner (2013) has shown that structured engineering design processes in combination with a commitment of the management are going to support the use of methods. And the adequate use of methods is leading to more successful products in shorter time.

2.4.1 Support for Getting from Society to Requirements

A large number of systematic methods like scenario technique, business planning and QFD are supporting the transfer between society with its needs or problems and the final definition of requirements. The successful use of the Kano-model as well as design for emotion in general is not possible without understanding the situation of and in the society, today and in future! A key point for engineering design is to understand the customer (decider, user etc.) next to all the technical, legal and organizational aspects. One of the proved methods is Storytelling including the role of a "story-keeper" to generate a continuous advocate of the user needs throughout the engineering design process (Michailidou et al. 2013).

This phase results in generating and defining the overall goals and the list of requirements. Creativity plays a major role in this phase, as all the stories and the requirements have to be oriented on future situations.

2.4.2 Support for Getting from Requirements to Solutions/Products

This is the classical approach for engineering design with a lot of helpful and also poor methods. In the view of a lot of practitioners as well as researchers this is the phase, in which creativity is required. Because of a number of restrictions like time and cost pressure all the creative steps should be goal oriented. Along the ongoing process this point is of increasing importance!

A large number of methods are claiming to support creativity especially in problem solving and finding/generating innovative solutions. A lot of empirical research has been done regarding efficiency and effectiveness. Most of the research is suffering under the problem of measuring the outcome!

At the end of this phase the documentation of the solution/product as input for production and purchasing as well as use, maintenance, recycling etc. is available.

2.4.3 Support for Getting from Solutions/Products to Society

Engineering design has to follow up the acceptance and use of solutions/products within society because of risk management reasons. But usually the further improvement, upgrading and learning for the generation of solutions/products are drivers.

Creativity in this phase is necessary to improve sales and services as well as doing the ongoing perfective maintenance in an adequate way. Generating the important lessons learned as far as these may be of importance in future business requires also some kind of creativity in selection.

2.5 Generic Procedure for Creating Solutions/Products

Within all the main-phases described in the above chapter the creation of specific results is required. The suggested procedure is similar to the problem solving procedure, as described in the MPM, the Munich Procedural Model (Lindemann 2009). Looking at the specific points of supporting creativity and based on experience out of a large number of case studies there are a number of steps that usually occur:

1. Recognize a situation that requires attention beyond routine (similar to (Badke-Schaub and Frankenberger 2004))! This may lead to a specific kind of activity, which could benefit from creativity.
2. Understand the situation/problem up to a sufficient level based on questionnaires, cause-effect analysis, and discussion (also with ourselves), stories etc.! The context information, important interdependencies and an adequate level of transparency are required! Because of required experience to judge about the "right" level of information on one side and the ambivalence of experience itself on the other side usually iterations will be necessary. Different perspectives like views of stakeholders, competitors, legislation, customer/user may support the overall understanding.

3. Describe the situation/problem on a more abstract level, as far as required and in a direction that may support further action! The degree of abstraction is dependent on the situation/problem and on the experience of the engineering designer.
4. Find analogies! There are a lot of possibilities to find or generate analogies like searching, combining known knowledge-elements, looking at other domains to get inspired etc. A lot of well-known methods like the morphological box, biomimetics, synectics or elements out of TRIZ (TIPS) (Pahl et al. 2007; Lindemann 2009) are supporting this task. Using these methods requires some training and in case of working in a team an experienced moderation is required.
5. Concretize found analogies in the sense of solutions and adapt these to the situation, the problem! This again is a demanding step, which requires the ability of open minded thinking. A lot of experience with regard to solution may hinder, experience in generating new solutions may help.
6. Evaluate the findings/results! The above defined requirements are the first point to be checked. An analysis of risks and chances, development efforts and time, availability of necessary resources may be additional aspects of an evaluation.
7. Decide!

Iterations within this procedure are necessary and helpful, as long as the number is kept on a small level. Teamwork may help in a few of these steps (mainly "Understand the situation/problem" and "Evaluate the findings/results") and it may hinder progress in others (mainly "Describe the situation/problem...", "Find analogies" and "Decide").

The "Concretize found analogies" seems to be the most demanding step with hardly any systematic support. All the other steps are overall well supported by systematic methods.

Novices may take more time and they need support during evaluation steps. They have a good chance to bridge the gap of missing experience by using the systematic way intensively.

Experienced practitioners are often hindered by mental blocks because of their experience. They have a good chance to bridge the gap based on these mental blocks by using the systematic way intensively.

2.6 Conclusion

Research in engineering design should address the points of teamwork versus individual work with regard to its basics as well as to industrial practice. Difficulties as measuring the output in terms of quality and the possibility of repeating experiments have to be addressed.

Research in engineering design should also address the point of concretizing found analogies in the sense of solutions and adaptation of these to the given situation/problem.

Heymann (2005) cited Hubka 1976: "Design is a demanding creative task, which should not be seen as art, but as scientific work." and Franke 1999: "Design is not a science."

Within a journal of a community of university professors Max Weber (sociologist from 1864 to 1920) was cited that "only based on hard work the intuition is prepared" (Forschung und Lehre 2013).

Coming back to creativity:

We will be able to generate/create successful solutions/products, if we start and end our processes with social motives/society and follow systematic ways of dealing with our tasks.

Up to now we have not understood all the mechanisms of thinking, but we are using and supporting our thinking capabilities. Up to now we have not understood all the mechanisms of creativity, but we are using and supporting or creative capabilities!

References

Badke-Schaub P, Frankenberger E (2004) Management Kritischer Situationen. Springer, Berlin

Erharter WA (2012) Kreativität gibt es nicht. Redline, München

Forschung und Lehre (2013) DHV, Bonn p 611, ISSN 0945–5604 8/13

Furnham A (2000) The brainstorming myth. Bus Strat Rev 11(4):21–28

Graner M (2013) Der Einsatz von Methoden in Produktentwicklungsprojekten. Springer Gabler, Wiesbaden

Heymann M (2005) "Kunst" und Wissenschaft in der Technik des 20. Jahrhunderts, Chronos, Zürich

Higgins JM (1994) 101 creative problem solving techniques. New Management Publishing Company, Winter Park

Lindemann U (2009) Methoden der Produktentwicklung. Springer, Berlin

Michailidou I, von Saucken C, Lindemann U (2013) How to create a user experience story. In: Marcus A (ed) Design, user experience, and usability. Design philosophy, methods, and tools, lecture notes in computer science, vol 8012. Springer, Berlin, pp 554–563

Pahl G, Beitz W, Feldhusen J, Grote KH, Wallace K, Blessing L (2007) Engineering design: a systematic approach. Springer, London

Chapter 3
Sign, Design, Communication

Michio Ito

Abstract In this article, we use communication as a point of reference to discuss a variant of the sign or mark that design produces. Using the works of French philosopher Jacques Derrida, we focus on the role of communication in design work. Derrida's analysis is in conflict with ordinary views, in that does not convey an identifiable meaning or ideas between some people. The sign and design in general must be able to function in the absence of the sender, the receiver, and the context of production (i.e., they must not lose their function in different settings), and there is always the possibility of transformation (iterability). This implies the possibility that transformation is always and necessarily inscribed in the functioning or functional structure of the sign and the design, and repeatability as iterability can coexist with originality. On the other hand, the sign and the design also have the capability to communicate, primarily as a means of communication. As a secondary effect, social communication is enhanced, usually based on the conventions used in the design. However, because social communications are inherently limited and iterable, design needs to consider the possibilities of iterability.

Keywords Communication • Iterability • Sign

3.1 Introduction

Our living sphere includes a plethora of products that are industrially, technically or artistically designed. However, when we consider a product's 'design', we immediately encounter difficulty. The word '*design*' has enormous number of meanings and use, in that it can be considered from a point of view of science and technology, engineering, art, society, problem-solving, risk, and so on.

Can design also be discussed philosophically?

The arguments about architecture design first arose in the 1960s and lasted until the 1980s, with most discussions revolving around how to criticize modern

---

M. Ito (✉)
Tokyo University of Agriculture and Technology, Koganei-shi, Tokyo, Japan
e-mail: petrus@cc.tuat.ac.jp

© Springer Japan 2015

T. Taura (ed.), *Principia Designae – Pre-Design, Design, and Post-Design*,
DOI 10.1007/978-4-431-54403-6_3

architecture and how to evaluate postmodern architecture. By criticizing the universalism and functionalism of modern architecture, the arguments moved toward a critique of 'postmodernism' in general.

Philosophical arguments such as those posed by French philosopher Jean-François Lyotard (1924–1998), focused on how the status of knowledge has changed in postmodern society: "The old principle that the acquisition of knowledge is indissociable from the training (*Bildung*) of minds, or even of individuals, is becoming obsolete and will become ever more so" (Lyotar 1979). Knowledge is and will be produced in order to be sold. And "the relationships of the suppliers and users of knowledge to the knowledge they supply and use is now tending, and will increasingly tend, to assume the form already taken by the relationship of commodity producers and consumers to the commodities they produce and consume—that is, the form of value" (Lyotar 1979, p. 14). Legitimation of knowledge produced during the Enlightenment was tied to "meta-narratives," or grand narratives that make ethical and political prescriptions for society, and generally regulate decision-making. Therefore, in one sense, postmodernism can be seen as incredulity toward meta-narratives.

Discussions around modern or postmodern architecture went far beyond design itself and tapped into theories about postmodern society.

How architecture design can be explored within philosophical discourse?

How can we discuss design philosophically?

3.2 Design and the History of Ideas

The word *design* derives from the Latin word '*desiganre*'. '*Signare*' means '*signum*', sign or mark, so 'designare' means to mark out, or to make a signum (sign) or a mark. What and how do we make a sign or a mark? And to what? Here we can say design (design activity) is primarily intended to make a sign.

'*Designare*' gave birth to the Italian word '*disegno*', which became one of the major concepts used by Renaissance artists. Fine arts, such as painting, sculpture and architect—there was no word corresponding to the meaning in those days—were said to be based on discegno.

Giorgio Vasari (1511–1574) defines the 'art of *disegno*' as painting, sculpture, and architect. Before Renaissance period painters, architectures and sculptors had belonged to the 'Guild', which was an association of artisans or merchants who controlled the practice of their craft in a particular town. However, Guild members became conscious of the difference between mechanical arts and visual or fine arts. Giorgio Vasari's *Le Vite de' più eccellenti pittori, scultori, e architettori da Cimabue insino a' tempi nostri* [The Lives of the Most Excellent Italian Painters, Sculptors, and Architects, from Cimabue to Our Times] (1550) was one of the first treatises that made this distinction (i.e., that the artiste is different from the artisan). Under his influence, the L'Accademia delle Arti del Disegno [Academy of the Arts of Drawing] was founded by Cosimo I de' Medici in 1563. The academy represented artistic independence from the medieval guild.

With the rise of Aesthetics as a science in the eighteenth century, philosophical arguments about the difference between the mechanical arts (*mecahnishche* Kunst) and fine arts (*schöne* Kunst) emerged again. Immanuel Kant (1724–1804), in his *Kritik der Urteilskraft* [Critique of Judgment] tries to classify "Kunst (art)". If art, Kant argues, is adequate to the cognition of a possible object and performs the actions requisite merely in order to make it actual, it is called *mechanical* art. If art has for its immediate design the feeling of pleasure, it is called *aesthetic* art. In addition, if pleasure universally accompanies the sensation of the art, it is called *pleasant* art (e.g., entertainment). If pleasure is a primary feature during cognition, a work is known as *beautiful* art (*schöne* Kunst). Beautiful art does not follow scientific cognition, but extends cognition with free imagination to form an "aesthetic idea." This classification was adopted as a standard for arts and design (Kant 1790, § 44)

On the other hand, the word '*disegno*' was translated in the seventeenth century into the French words '*dessin*' (drawing) and '*dessein*' (intention or project). The two semantic dimensions of the word '*disegno*' were separated. Design, as far as it means '*dessein*', which implies intentionality and spirituality, fulfills this condition of art.

Following aesthetics, modern art identified itself with the original and creative talent of the author and, at the same time, excluded natural beauty without spirituality. Therefore, traditional aesthetics tends to exclude the conception of natural beauty. In his *Aesthetics*, Georg Wilhelm F. Hegel (1770–1831) says that aesthetics is "the science which is meant deals not with the beautiful as such but simply with the beauty of art... Spirit and its artistic beauty stands *higher* than natural beauty". Therefore, aesthetics refers to "the philosophy of art and, more specifically, the philosophy of fine art" (Hegel 1970, Einleitung p. 13–14).

Design separated from '*dessein*' means merely '*dessin*', which references the possibility of formal drawing, which is a building block of artistic expression. This formal design (*dessin*), however, will lead to produce mechanical design.

By the end of the nineteenth century, with the birth of formalistic aesthetics, arts and crafts movements, and modern technology, the two semantic notions could again be united. Such circumstances gave rise to product and industrial design. In fact, the English word '*design*', as used today, emerged during the early days of industrial capitalism. The word includes both meanings (drawing and intention) and is usually used directly (i.e., not in translation) in non-English-speaking countries (Kennichi 2004). In this sense, indeed, a rethinking of the history of design has led to a rethinking of 'modernism' and 'postmodernism'.

3.3 Visible and Invisible Design

As detailed above, *design* means to mark out. Through marking or drawing, we can visualize the form of an object and grasp its relation to a designer. In this way, we can say that we come to have a point of view about a design. However, there is no single point of view. German philosopher Friedrich Nietzsche named this '*perspectivism*': "In so far as the word 'knowledge' has any meaning, the world is knowable; but it is interpretable otherwise, it has no meaning behind it, but countless meanings—Perspectivism" (Nietzche 1906, § 481)

Does the object that design marks out have an outline? Traditional metaphysics has supposed that objects or entities have an identity. Therefore, objects do indeed have an outline that separates themselves from other like objects. We can consider this view as metaphysics of individuality and simple directness. The word *metaphysics* means in this case a theory in general that classifies and institutes a hierarchy of values; individuality and simple directness are prior to anything, and totality or complexity and indirectness are secondary.

If we suppose this metaphysics, only one design would seem to outline the 'real' world, with scientific empiricism or technical utility used to justify the design. However, this is one of many perspectives. Design as art hearkens back to an alternative perception of the world, and objects in that world are a priori. Thus, Impressionism destroyed "a prosaic conception of the line".

In the history of painting, indeed:

> There has been, for example, a prosaic conception of the line as a positive attribute and property of the object in itself. Thus, it is the outer contour of the apple or the border between the plowed field and the meadow, considered as present the world, such that, guided by points taken from the real world, the pencil or brush would only have to pass over them. (Merleau-Ponty 1961, p. 72)

However, modern art or painting has come to deny this edict. Rather, there should be no visible lines. Lines are "always between or behind whatever we fix our eyes upon; they are indicated, implicated, and even very imperiously demanded by things, but they themselves are not things." The lines imitate no longer the visible. They render visible. They are "the blueprint of a genesis of things" (Merleau-Ponty 1961, p. 73). Thus, merely *dessin* has also spiritual activity (i.e., we cannot think about design without spirituality). Is there truth to this claim?

We can argue otherwise: when an object outlined by marks or lines emerges, the marks or lines that delineate the object disappear and we can see only their traces. Occasionally, we cannot even see the traces. The limit or the border that forms the essence of the marks or the lines is not an entity but the difference itself. That is, the marks or the lines, which render something visible, cannot be seen as themselves. The condition of visibility itself includes invisibility. Following Plato, traditional metaphysics has tried to interpret spirituality based on an optical model. Plato's "δέα, idea" is originally derived from the Greek verb '*idein* (see)' and considers the meaning of a form that is seen; to understand is to see. Therefore, to understand spirit requires visualization. If visibility includes invisibility, might spirituality include in itself the impossibility of spirituality?

3.4 Design Relationship

Design does not stand-alone: it includes many facets that give rise to communication and interpretation. We often design something using former products or processes as a model. Moreover, designers often quote former designs and design plans. Design stands between the former and the following, and the real and the imagined, which may include projected but uncompleted designs that have never been evaluated. We can call this aspect of design 'inter-design', which relates itself to 'pre-design'. In the pre-design phase, communication occurs that may drive actors toward the conception of a new design.

A design may be criticized or interpreted during its conception. We call this 'meta-design', which has two phases. First, if a design is simply situated in relationship to a former product, it is considered as a process of 'post-design'. Conversely, if the design produces another inter-design series in another system, it becomes 'proper meta-design'. A meta-design can be a transformed referential design—partially or totally, sincerely or playfully (e.g., pastiche, parody), and so on. Since the avant-garde movement, contemporary art defines itself based neither on the imitation of nature nor on the beauty that justifies traditionally beautiful art. It refers not to its own content, but to its form and preceding works. As a result, recent artwork has become more and more self-reflexive, reflecting an infinite meta-design character. Such self-reflexive repetition/iteration of art will form nonlinear, disseminating time series.

Design tends to leave marks that classify itself. For example, an artistic design often includes the author's name or a distinct title, or sometimes even an intentional "Untitled." An untitled work is often conceived of with the purpose of escaping the power of language and fully developing its own form of expression, so that the idea of being "untitled" functions as the title itself. On the other hand, technological design sometimes includes only the brand or product name. These considerations fall under what we call the "para-design" phase, which also has two facets: outer space and inner space. The former includes, for example, interviews or reviews, and the latter includes titles and the author's name (Fig. 3.1). Both facets tread a path toward social communication (Genette 1982).

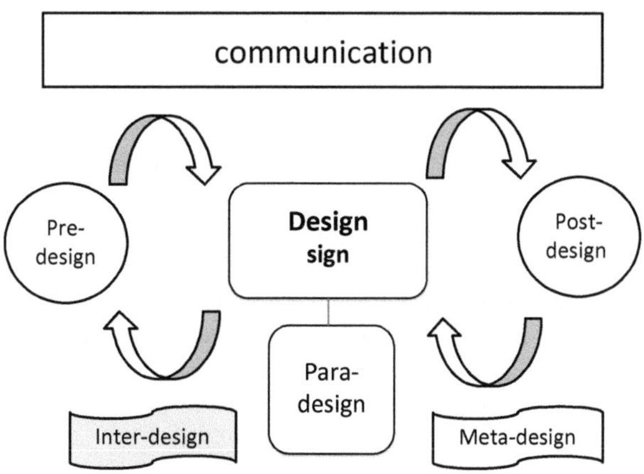

Fig. 3.1 Design relationship and communication

3.5 Communication and Iterability

When designing a product, we communicate around a series of specifications. In other words, the design gives rise to communication. If we consider design as a process of communication, does the meaning or function of a design reside in the same sphere as communication itself?

Then, what is communication? Communication conveys meaning; that is, communication is the transmission or the conveyance of content or information. Communication is often thought of as the exchange of information between two or more individuals based on a common code or communication ability. In this process, the meaning is appropriated through a system that conveys the intention of both actors. Therefore, the identity of meaning is supported by the 'code' or 'intentionality' of a speaker or a sender. However, is it "certain that to the word communication corresponds a concept that is unique, univocal, rigorously controllable, and transmittable: in a word, communicable?" (Derrida 1988, p. 1)

Moreover, we often say that communication takes place within a certain 'context'. "But are the conditions of a context ever absolutely determinable?" "Is there a rigorous and scientific concept of context?" (Derrida 1988, p. 3)

A French philosopher Jacques Derrida (1930–2004) offers a series of philosophical conceptions related to communication (which he denied as concepts): déconstruction (deconstruction), différance (deferring), dissémination (dissemination), supplément (supplement), écriture (writing), and so on.

He aimed all these of concepts at the "métaphysique de presence" (metaphysics of presence). In Derrida's conception,

> The history of metaphysics, like the history of the West . . .is the determination of Being as the presence in all senses of this word. It could be shown that all the names related to fundamentals, to principles, or to the center have always designated an invariable presence—eidos, arche, telos, energeia, ousia (essence, existence, substance, subject) aletheia, transcendentality, consciousness, God, man, and so forth. (Derrida 1967a, pp. 410–411)

We can add more: intention, meaning, purpose, ends, plan, or the presence to sense (empiricism), the presence to intelligence (rationalism), and the presence from exteriors to interiors, and so on.

According to Derrida, in the history of Western philosophy oral communication predates writing. In Plato's work, truth or *logos* is present in spoken language; the presence of intentions, as well as context, leads a listener to understand the meaning of a speaker and ostensibly avoid misunderstanding. Therefore, spoken language is the model or archetype of a language. On the other hand, written language is its supplement, and in the case of the absence of a speaker, it represents the meaning that was originally presented in and by spoken language.

Contrary to such a view, Derrida asks whether 'writings' invade the pure presence of spoken language and threatens its meaning, while the metaphysics of presence conceal it. Derrida criticizes the classical concept of writing in terms of communication of intended meaning. If a communication transmits a certain meaning, it presupposes:

1. What the sender has to communicate is his or her 'thoughts', 'ideas', 'representations', and 'references'. The contents of communication are ideas, thoughts, information, signified contents, and so on. The intention of the sender precedes and governs communication processes that transport the 'idea'—the signified content. Thus, the same meaning in communication is based on intention.
2. When a communication is completed, the ideas based on the intention of the sender are present to the receiver through the sign that re-presents thought, ideas, or meaning. Thus, in an ideal exchange the very same idea that is present with the sender will be re-presented with the receiver. Anything that prevents this sameness between presentation and re-presentation is 'noises' improper to communication excluded. The receiver interprets messages by the sender on the basis that of contextuality and intentionality determines meaning.
3. The presence of meaning is fully enriched in spoken language, in which the presence of speakers, intentions, and context lead to an understanding of meaning. In other words, spoken language is the model language.
4. Supposing a homogenous space, and in a continuous manner, people invent particular means of communication, writing, and so on. Articulated language has come to "supplant [suppléer]" the language of action and then, writing supplants articulated language. The birth and progress of writing will follow in a line that is direct, simple, and continuous (Derrida 1988, p. 4). Re-presentation regularly supplants (*supplée*) presence. Therefore, we can translate the same meaning of presence into written language.

The word '*supplant*' implies the double notion of supplanting, replacing, and supplementing, bringing to completion, remedying.

> For the concept of the supplement harbors ... within itself two significations whose cohabitation is as strange as it is necessary. The supplement adds itself, it is a surplus, a plenitude enriching another plenitude, the *fullest measure* of presence. It cumulates and accumulates presence. It is thus that art, *technè*, image, representation, convention, etc., come as supplements to nature and are rich with this entire cumulating function ... But the supplement supplements. It adds only to replace. It intervenes or insinuates itself in-the-place-of; if it fills, it is as if one fills a void. If it represents and makes an image, it is by the anterior default of a presence. Compensatory [*suppléant*] and vicarious, the supplement is an adjunct, a subaltern instance which *takes-(the)-place* [*tient-lieu*]. (Derrida 1967b, p. 208)

As seen above, writing has been thought of as a means of communication and extension of oral or gestural communication. To say that writing extends the field and the power of oral or gestural communication presupposes a sort of homogeneous space of communication. In this homogeneous and linear process,

> writing will never have the slightest effect on either the structure or the contents of the meaning (the ideas) that it is supposed to transmit. The same content, formerly communicated by gestures and sounds, will henceforth be transmitted by writing, by successively different modes of notation, from pictographic writing to alphabetic writing. (Derrida 1988, p. 4)

Therefore, we can say that the written communication has representational character. *Representation* means, in this case, to supplement the presence of a thought or idea and represent or express meaning in the absence of oral communication.

Indeed, a "written sign is proffered in the absence of the receiver." We can read text in the absence of an author, of course, so we can read archeological signs, marks, or traces and continue to read the text of Plato, perhaps with some possibility of misunderstanding. Is this absence not merely a distant presence or re-presentable presence? Arguably, it is not. In order for my "written communication" to retain its function as writing (i.e., its readability), it must remain legible despite the absolute disappearance of any receiver, determined in general. To be able to read or interpret signs, marks, or language, they must be repeatable in the absence of anyone.

This repeatability, Derrida says it in the different word '*iterability*'.

> Such iterability-(*iter*, again, probably comes from *itara*, other in Sanskrit, and everything that follows can be read as the working out of the logic that ties repetition to alterity) structures the mark of writing itself, no matter what particular type of writing is involved. (Derrida 1988, p. 7)

The absence of a receiver and repeatability or iterability leads to the absence of a sender or producer. "To write is to produce a mark that will constitute a sort of machine which is productive in turn, and which my future disappearance will not, in principle, hinder in its functioning, offering things and itself to be read and to be rewritten" (Derrida 1988, p. 6). For writing to be writing, it must continue to "act" and to be readable even when what is called the author of the writing no longer answers for what he has written—for what he seems to have signed. Moreover, if writing is to be readable in the absence of a receiver and a sender, it must be readable in the absence of the context in which that the writing was formerly inscribed: "A written sign carries with it a force that breaks with its context" (Derrida 1988, p. 9).

Because written signs have repeatability, they can always be detached from the context "in which it is inserted or given without causing it to lose all possibility of functioning." We can always recognize other possibilities in them by inscribing them or grafting them onto another context. "No context can entirely enclose it. Nor any code" (Derrida 1988, p. 65). We can quote written signs in any context. The written signs grafted onto another context can still represent a meaning, which has the both possibility of repeatability of the same meaning and of the transformed meaning (iterability).

This rupture from context allows written signs to serve as separate 'references'. Their function and readability must work in the absence of a receiver, sender, context, and the intentional signified references. This absence is inherent in the structure itself of a sign or language, thus the possibility of iterability must be taken into account if we are to construct an evolving communication theory. Therefore, this possibility is always inscribed in the functioning or the functional structure of a mark. This character will be, in particular, seen in the signature:

> By definition, a written signature implies the actual or empirical non presence of the signer. However, it will be claimed, the signature also marks and retains his having-been present in a past now or present which will remain a future now or present. (Derrida 1988, p. 20)

When someone signs their name to a document, it is taken as an original, specific fact that governs other signatures. A banker may refer a signature and confirm it with past signatures. So, even in the absence of a physical entity, a signature is recognized as a representation of our will and intention. Nevertheless, the only course of action is to reference the present sign with the original estimated one:

> Effects of signature are the most common thing in the world, but the condition of possibility of those effects is simultaneously, once again, the condition of their impossibility, of the impossibility of their rigorous purity. In order to function, that is, to be readable, a signature must have a repeatable, iterable, imitable form; it must be able to be detached from the present and singular intention of its production. It is its sameness which, by corrupting its identity and its singularity, divides its seal [*sceau*]. (Derrida 1988, p. 20)

A signature that takes place only once has no true function and may not even be admitted as a signature. A signature, as long as we accept its validity, has the possibility of plural occurrences. In other words, a signature may be viewed as an event that divides itself logically, as in the case of other biometrics. The meaning or the function of the signature is not prior to words or language, just like what he or she says or means (*vouloir dire*)—the meaning is the effect of the signature.

The identifiable, repeatable and at the same time iterable character may also be confirmed by spoken language: "Through empirical variations of tone, voice, etc., possibly of a certain accent, for example, we must be able to recognize the identity, roughly speaking, of a signifying form" (Derrida 1988, p. 10).

On the other hand, a strong and influential tradition of Western philosophy still supports the identity of meaning. For example,

Aristotle says that

> by a 'sense' is meant what has the power of receiving into itself the sensible forms of things without the matter. This must be conceived of as taking place in the way in which a piece of wax takes on the impress of a signet-ring without the iron or gold; we say that what produces the impression is a signet of bronze or gold, but its particular metallic constitution makes no difference: in a similar way the sense is affected by what is colored or flavored or sounding, but it is indifferent what in each case the substance is. (Aristotle 1931, Part 12)

According to Aristotle, things or objects are combinations of form and matter or material. We receive a form without material. Sense organs, as far as they are attached to a body, are entangled material, but in the inner, mental sphere, we can receive the form of the object apart from materials. During speech, we receive the identical form of meaning apart from changing materials such as tone of voice, inflection, etc. We can always transit from the outside into the inside to eliminate materiality, and come to take a sound as an articulated identifiable voice not as a noise. Modern German philosopher Edmund Husserl names this transition from a sound to articulated voice, or from an ink stain to writing, as the "meaning giving intentionality of a consciousness" (Husserl 1984, Section I. 'Expression and Meaning', § 23).

Alternatively, according to linguist Ferdinand de Saussure (1857–1913), it is possible to receive a phoneme apart from the voice that occurs in space and time.

After elimination of materiality, which is inherent to the outside object and the body, there remains the 'form' of inner sphere. The concept of 'form' comes originally from Plato's concept of '*idea*' or '*eidos*' (essence), which guarantees that identity is interpreted in modern philosophy as the identity of objects, consciousness, or spirit. Nevertheless, we can read a sign without presupposing the doubtful concept of the 'form' as idea or code, and as long as a sign is still readable, the form must be taken as a repeatable and iterable one that has materiality.

The word '*writing* (*écriture*)' used by Derrida does not always mean writing in contrast to spoken language. The oppositions between speech and writing (intelligible and sensible, soul and body) seem to have persisted throughout the history of Western philosophy. The former, the superior term, belongs to presence and the logos; the latter, the inferior term, serves to define its status and mark a fall. It is not important to reverse this hierarchical opposition. It is not a simple valorization of writing over speech. Indeed, "the usual notion of writing in the narrow sense does contain the elements of the structure of writing in general: the absence of the 'author' and of the 'subject-matter,' interpretability, the deployment of a space and a time that is not 'its own.' We 'recognize' all this in writing in the narrow sense and 'repress' it" (Derrida 1967b, Gayatri Chakravorty Spivack, translator's preface. p. lxx). Therefore, we ignore that everything else is also inhabited by the structure of writing as *écriture*.

Indeed, Derrida admits this repeatability and iterability in our experience. The marks that constitute experience construct "the network of effacement and of difference, of units of iterability, which are separable from their internal and external context and also from themselves". (Derrida 1988, p. 10)

A work of fine art seems to have an original, specific identity. Our experience also seems to confirm it. But how? Our artistic experience—could it be possible only through the cognition of the identity and the aesthetic form of the work? This puts the cart before the horse. The identity in our aesthetic experience does not depend on the form presupposed by transcendental metaphysics. Without a repeatable and iterable aesthetic experience, we are capable of discerning the original piece and a fake. Rather, such an experience will constitute the identity with iterability: Something that carries within itself the trace of a perennial alterity: the structure of the psyche, the structure of the sign. To this structure, Derrida gives the name 'writing' (*écriture*).

The structure of the sign, of experience, and of text has logically, even if not in a fact, alterity or iterability. We usually think that originality or specialty comes from a specific talent or fortuity and consists of only one event. However, originality can and does coexist with repeatability as iterability. To write or design is to produce a mark that constitutes a sort of machine that repeats the same action, but then in turn repeatedly produces something different, positive or negative.

> Let us not forget that "iterability" does not signify simply, . . . repeatability of the same, but rather alterability of this same idealized in the singularity of the event, for instance, in this or that speech act. It entails the necessity of thinking at once both the rule and the event, concept and singularity. There is thus a reapplication (without transparent self-reflection and without pure self-identity) of the principle of iterability to a concept of iterability that is

> never pure. here is no idealization without (identificatory) iterability; but for the same reason, for reasons of (altering) iterability, there is no idealization that keeps itself pure, safe from all contamination. The concept of iterability is this Singular concept that renders possible the silhouette of ideality, and hence of the concept, and hence of all distinction, of all conceptual opposition. (Derrida 1990, 'Afterword' p. 119)

Then, "communication, if we retain that word, is not the means of transference of meaning, [but] the exchange of intentions and meanings, discourse and the communication of consciousnesses." As a result, communication as writing would no longer be "one species of communication and all the concepts to whose generality writing had been subordinated (including the concept itself qua meaning, idea or grasp of meaning and of idea, the concept of communication, of the sign, etc.) would be denied" (Derrida 1988, p. 20).

3.6 Force, Communication and Design

How can we begin to think in this way? When we trace a sign or a mark, the possibilities of iterable communication will follow. If we read an animal footprint and look for a way of follow them, there is the possibility of failure (iterability)—a footprint or a signature can be a fake. Where there are intentional or unintentional (natural) marks, iterable communication takes place. Or rather, a sign or a mark forces us to read as long as we have repeatable or iterable experiences. Any sign or mark has this power. Can this structure correspond to design? When we design something, the possibilities of iterable communication take place. Therefore, the design will always be detached from the context and grafted onto other contexts—it can be transferred anywhere and may unexpectedly function, which may bring disaster or new flourishing.

In the second phase, a design can have the force or power to communicate, similar to the force of language. This force can be viewed as an "illocutionary act," a term John L. Austin introduced in his theoretical book about speech acts. Austin identifies three types of speech acts: *a locutionary act*, the actual performance of utterance; *an illocutionary act*, or the '*illocutionary force*' of the utterance (this force of a socially valid utterance is based on conventions); and *a perlocutionary act*, which has an actual effect, such as persuading, convincing, or otherwise getting someone to do something, whether intended or not (Austin 1962).

A typical example of an illocutionary act is promising or ordering someone. When we say "Get away!" we order and do not describe a fact. Through saying to someone "Get away!", we order, that is, "by saying something, we do something" (Austin 1962, Lecture 9). The utterance of the sentence has force to bring about the action or the performance that the sentence describes. Other examples include: "I name this ship the Queen Elizabeth," "I bet you six pence it will rain tomorrow," or "I promise to pay you back." These are performative utterances, which have the power of illocutionary act. In addition, Austin regards not only such performative sentences, but also the sentence that describes the fact (the constative utterance) as having such force.

Derrida claims Austin that he appears to consider speech acts only as acts of so-called ordinary communication so that he seeks to the limit of the context where the original, intended, serious and normal, but not abnormal or parasitic meaning will be expressed and this force will be worked. Austin says that "a performative utterance will, for example, be in a peculiar way hollow or void if said by an actor on the stage, or if introduced in a poem, or spoken in soliloquy" (Derrida 1990, p. 16, Austin 1962, pp. 21–22). In those cases, language is used "not seriously". The proper force of utterance occurs under the *presence* of a proper context or condition.

Derrida ask; "could a performative utterance succeed ... if the formula I pronounce in order to open a meeting, launch a ship or a marriage were not identifiable as conforming with an iterable model, if it were not then identifiable in some way as a 'citation?'" (Derrida 1988, p. 18). Forgery of the signature is always possible; the possibility of transgression is always inscribed in speech acts. Derrida points out that even though Austin recognizes the possibilities of the risk or exposure to infelicities, he does not ponder a possible risk, and considers solely the context and the conventionality constituting the context and aims the presence of a total context to speech act. Austin seems also to share the tradition of the metaphysics of the presence.

At the same time, Derrida appreciates Austin: "Austin's notions of illocution and perlocution do not designate the transference or passage of a thought-content, but, in some way, the communication of an original movement" as "an operation and the production of an effect". Performative communication "would be tantamount to communicating a force through the impetus [impulsion] of a mark". A performative utterance "does not describe something that exists outside of language and prior to it. It produces or transforms a situation, it effects" (Derrida 1988, p. 13).

Of course, a sign or a mark is involved to iterability as a primary condition, while it gives rise to so-called communication as a secondary effect; an effect is subordinated to a cause, emerging from the system but never controlling the whole system. The intention of the design, its function or intended purpose (e.g., to solve a problem) and its cause or the motive of the design (e.g., social, political, technological demands), are, therefore, secondary effects for the design.

As a secondary effect, any utterance has a force to build relations between people. By saying "I promise," a social relation is built up and responsibility to the utterance will occur. Austin says that conventions constituting the context support it. However, if the utterance has an iterable character and the context can be variable, and the utterance works without conventions, the responsibility of the speaker is limited because he or she cannot control every variable.

Design also seems to be performative. Artistic design, as well as industrial design and engineering design, does not merely describe a fact. Its function has the power to produce or transform a situation. In other words, the design has some influence in forcing us to act, produce, or communicate—it reifies social communication and, because of agreement during the development phase, ensures that the design and communication will probably circulate. However, the social communication reveals itself to us only in a familiar form (e.g., as a social or technological demand). "What is familiarly known is not properly known, just for the reason that it is familiar. When engaged in the process of knowing, it is the commonest form of self-deception and a deception of other people as well" (Hegel 1986, Vorrede p. 35). Moreover, other communication process may go out of sight in names of function, purpose, or utility, which conceals and derives the communication process from the design. Sometimes scientific, technological, economic, or political terms govern the communication to the point that it excludes other forms.

We may regard a social system to survive as long as social communications continue. Social communications based on high technology aim to speed up the presence and re-presence of messages. However, if social communications have an iterable structure, those communications convey alterity and they may be delayed or not reach anyone; they may not communicate. A design has, at its core, a structure of iterability; so do social communications. Therefore, for example, risk communications can also have iterability, and a design based on risk communication can logically fail to control for certain risks. As a result, when designing something, we also must design its capacity for social communication. And if social communication implies iterability, what we can do is to respond to multiplied communications in different contexts that we can never foresee or control, and design social communication that takes into account the impossibility of full control.

Responsibility means considering the possibility of how to respond to others. Responses do not always occur consciously (i.e., a response includes not only a subjective self-decision, but also a passive response). Does my responsibility end when and where the identical person, I, die or others die? Does it survive my death?

According to Kant, the condition of morality presumes three elements: free will, immortality of soul and God. Without free will, there is no responsibility. Of course, we design something with free will. In addition, our response to design and communication survive our death as writing, which includes the conception of design: it survives after the designer's death as heritage or successive social communication. Then, it may correspond to immortality of soul. God, as the absolute other to others and us, relates us beyond time and space. This relationship corresponds to social communication through design, for communication relates us to others beyond not only contemporary communities or nations, but also in the future or in the past. Therefore, design lets us to response all others. In other words, the design has a performative force to build a social communication in which we have responsibility.

3.7 A Grammatology of Design

> The sign cannot be taken as a homogeneous unit bridging an origin (referent) and an end (meaning), as 'semiology,' the study of signs, would have it. The sign must be studied 'under erasure,' always already inhabited by the trace of another sign which never appears as such. 'Semiology' must give place to 'grammatology.' (Derrida 1967b, Translator's preface. p. xl)

Therefore, the design as well as the sign must be studied under 'grammatology.'

1. The design (the designed mark-system) works apart from the intention or purpose of the designer, from the receiver, and from the context or the situation where formerly designed. It can function without any original motives or circumstances.
2. The design can break with every given context, engendering an infinity of new contexts in a manner that is illimitable. The context always opens to another and does not close—it cannot be wholly described.
3. The design has the possibilities of iterability. It has logical possibilities to transform into different functions from the original intention. It can function in another way, and any attempt to normalize it will fail because of the openness of the field. Designing is to sign or mark that which has an iterable structure. Originality can and does coexist with repeatability as iterability.
4. The design has power to produce or transform a context and to force us to read, act, communicate, or use. Conversely, the design is the reification of communication.
5. As a secondary effect, the design has a force to build a social relationship or communication in which we have responsibility. However, because of an iterable structure, the social communication does not reveal itself; rather, reveals only one facet of the design. The design should not conceal iterable function.

References

Aristotle (1931) On the Soul (translated by John Alexander). MIT Internet Classics Archive

Austin JL (1962, 1975) How to do things with words. Oxford University Press, Oxford

Derrida J (1967a) L'écriture et la différence, Éditions du seuil, Point, pp 410–411 (Writing and difference, translated, with an introduction and additional notes, by Alan B, Routledge, Kegan Paul, Ltd., London/New York)

Derrida J (1967b) De la Grammatologie, Les éditions de Minuit (Of Grammatology, translated with an preface by Gayatri Chakravorty Spivack)

Derrida J (1988) Signature event context [Signature événement context] p 20. In: Limited Inc. (trans: Samuel W, Jeffrey M) Northwestern University Press, Evanston, IL. Derrida J, *Limited Inc*, Galilée, 1990, (Derrida 1990). This article also in Derrida J (1972) Marges de la philosophie, Les éditions de minuit. Margins of Phhilosophy, translated by Alan Bass, The Harvest Press, 1982. 'Signature Event Context', the first English translation by Samuel Weber and Jeffrey Mehlman, appeared in volume 1 of the serial publication *Glyph* I in 1977

Genette G (1982) The prefixes 'inter-', 'meta-', and 'para-' are borrowed from text theory, especially, Palimpsestes. La Littérature au second degré (The Literature of second degree), Éditions du Seuil

Hegel GWF (1986) Phänomenoligie des Geistet 1807. Suhrkamp, Phenomenology of Spirit (translate by Miller AV, Oxford University Press, 1976)

Hegel GWH (1970) Vorlesung über Aesthetik, 1835–1838, Suhrkamp, Aesthetics, lectures on fine art (translated by T.M.Knox, Clarendon, 1975)

Husserl E (1984) Logische Untersuchungen [*Logical Investigations*,] Zweiter Band: Untersuchungen zur Phänomenologie und Theorie der Erkenntnis. Hrsg. von Ursula Panzer. Husserliana, XIX, Nijhoff, 1984

Kant I (1790) Kritik der Urteilskraft, section 44. Critique of judgment. English edition: Kant I (1914) Introduction and notes (trans: Bernard JH). Macmillan, London

Kennichi S (2004) Introduction to aesthetics, Tyuokouronshinsha

Lyotar J-F (1979) La condition postmoderne, Édition du Minuit. English edition: Lyotar J-F (1984) The postmodern condition (trans: Bennington G, Massumi B (1984). Manchester University Press

Merleau-Ponty M (1961) L'Œil et l'esprit, Gallimard, p 72. English edition: Merleau-Ponty M (1964) Eye and mind (trans: Carleton D, Eye and mind. In: James E (ed) The primacy of perception) Northwestern University Press, Evanston

Nietzche F (1906) Wille zur Macht (The will to power), Taschen-Ausgabe, Band IX, Naumann Verlag Leipzig

Part II
Culture and Cognition of Pre-design, Design, and Post-design

Chapter 4
A Sense of Design: The Embedded Motives of Nature, Culture, and Future

Yukari Nagai

Abstract To identify the meaning of the advanced design of the twenty-first century, this paper extracts embedded senses—nature, culture, and future—that activate design in society. To identify nature-related social motives that challenge design from a deeply emotional level, sustainability regarding extinct animals is first considered from a perspective that considers the whole environment. Second, the sense of culture is discussed in connection with designed products including abstract paintings. This shows the paradox of the self-organization system between humans and products. Openness is identified as the social motive from a cultural perspective. Third, a socially innovative design perspective is used to point out future subjects. This aspect shows reasons for the expectations of the deep design thinker who leads design discourse into social innovation from an ethical viewpoint, which are social motives for the sense of future. A discussion of these three aspects is used to determine the vital structure of the new meaning of design. Based on the discussion, characteristics of design creativity are presented while the role of academics in advanced design is explored.

Keywords Creativity • Deep design thinking • Design • Sense of culture • Sense of future • Sense of nature

4.1 Introduction

4.1.1 View of Advanced Design

The purpose of this chapter is to find a new meaning for creativity in advanced design. The meaning of creativity in advanced design is explained with an inside focus (namely, an "inner sense"), rather than the outside focus that commonly influences design process goals employed in definitions of conventional design (Taura and Nagai 2012). An inner sense is beyond perception; it involves a natural appreciation or ability to comprehend the external world. It is inferred that an inner

Y. Nagai (✉)
Japan Advanced Institute of Science and Technology, Ishikawa, Japan
e-mail: ynagai@jaist.ac.jp

© Springer Japan 2015

T. Taura (ed.), *Principia Designae – Pre-Design, Design, and Post-Design*,
DOI 10.1007/978-4-431-54403-6_4

sense activates intuition (or insight) as an underlying subconscious motivation. With regard to the relationship between a person and society, it is assumed that the resonation of an inner sense with a socially embedded sense is a core motivation for design. In other words, a socially embedded sense can trigger design from the inner sense of the subconscious mind. In this chapter, the notion of a "design sense" represents this resonance in organic activity.

To discuss these issues, we propose a paradigm of advanced design that involves phases of Pre-Design, Design, and Post-Design (Taura 2014). This paradigm can help us reach our intention of understanding the essential property of the Pre-Design phase and the social motive needed to change the "design discourse" of the future. Specifically, challenging actions by academics are encouraged. Some critical embedded problems of conventional design are extracted and replaced as social motives by considering conceptual roots.

By considering the relationships among the Pre-Design, Design, and Post-Design phases (Editorial Board of IJDCI 2013; Taura 2014), the state of society is determined from a "design creativity" perspective. Additionally, a mechanism to accelerate the motivation of design for advancing society can be identified by recognizing the internalized driving force of social motives. The priority should not be requests by industrial society for a definition of design. Rather, design should be understood as a creative cognitive process that characterizes humans. Nevertheless, this does not imply that industry is not important: industry is part of the social system, the meaning of advanced design must include the role of industry, and identifying advanced design as an autonomous system is an important step toward the future. In other words, industry is one organ of a holistic system organized with a future-minded sense of grand design.

4.1.2 Senses of Nature, Culture, and Future

Human creativity involves three basic senses—nature, culture, and future—which characterize humans. Those basic senses underlie human cognition. In other words, they are the embedded senses of humans and society represented as abstract concepts. For instance, image schemas are understood to represent an embedded, unconscious sense of nature (Johnson 1987; Lakoff 1987). To identify a new meaning of creativity that drives advanced design, this chapter focuses on the abstract concepts that represent the human senses of nature, culture, and future.

What is the sense of nature? This chapter views the sense of nature as a social motive stimulus that activates design from the human subconscious. Thus, the sense of nature is a typical example that explains the structure of the Pre-Design phase. The second section of this chapter demonstrates how intangible information in society is enhanced through deep emotions stirred by simple messages rather than by a large number of analytical reports concerning the environmental crisis (this issue is discussed in Sect. 4.2).

Accordingly, the sense of culture represents artifacts and their production, which is an important aspect of design creativity phenomena. The sense of culture dynamically interacts with human activities, habits, media, artifacts, and so on, which are formed through human communication and lifestyle processes. Therefore, the sense of culture involves the Design phase, which is driven by social motives formed during the Pre-Design phase. To discuss this connection, the meaning of fine art, considering the structure of creative processes, will be reviewed as an example (as described in Sect. 4.3).

The sense of future is an extremely abstract concept (Taura and Nagai 2012) that is related to design creativity competence. It represents the complex social cognition of change and uncertain issues. It is assumed that the connection between the Post-Design phase and Pre-Design phase regulates or constrains the Design phase. As such, it is expected that morality and ethics will appear as social motives (stemming from the sense of future) during the course of this discussion (as described in Sect. 4.4).

Social motives for advanced design are discussed to determine the senses described earlier. Each sense—nature, culture, and future—suggests a particular value that is embedded in society and forms representative concepts, or "meanings." Figure 4.1 shows the annotated directions of senses during the three phases of the design model.

The relationships between the phases of design in regard to oriented senses, social motives, and embedded knowledge—which are the intrinsic criteria of creativity that also serve as internalized driving forces for various expected types of innovative design—are classified in Table 4.1.

Mature design creativity in an advanced society efficiently encourages connections between the design phases through discernment and insight.

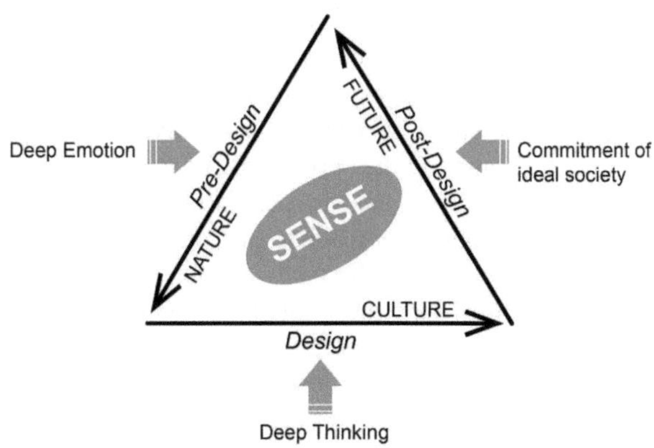

Fig. 4.1 Relationships between the senses of design during the three advanced design phases

Table 4.1 Relationships among phases of design, senses, motivations, social motives, and expected innovation types

Phase	Sense	Initialized driving force of creativity (motivation)	Social motive	Type of innovation
Pre-Design	Nature	Ego, Needs, Goal-setting	Sustainability	Counter-production
Design	Culture	Self-organization, Originality	Openness	Social networking
Post-Design	Future	Deep emotion, Deep thinking	Commitment	Ethics, Discourse

4.1.3 Meaning of Design Creativity

To develop the discussion on "design creativity" regarding a new design paradigm—which overlaps with the scope of this book—that includes Pre-Design, Design, and Post-Design phases, the meaning of "design creativity" must be carefully defined. This chapter employs the term "design creativity" to express, from a design perspective, the creative features of human cognitive processes and the creative process in society.

Theories on past and present uses of artifacts (including art) are reviewed considering the relationship between the senses. Additionally, my recent short message on the meaning of design for the future highlighted critical issues with regard to design creativity and innovative research (Editorial Board of IJDCI 2013). These issues are expanded in this chapter by emphasizing critical differences between conventional and advanced design. A main claim of my short message was the importance of a renewed design theme (motive), which should be generated to change the meaning of design. This chapter develops that claim by considering the relationship between socially and personally motivated design in order to expand ideas (inner criteria) with regard to future design. Based on this discussion, the latter part of this chapter takes on a future-oriented perspective.

There are also large gaps between the Design and Post-Design phases. The unfortunate situation that has caused conventional design is discussed in the next section. Recognizing the seriousness of the situation will lead to great changes in the social motives of the Pre-Design phase from a nature perspective. Following this, a psychological approach to the cognitive link between happiness and creative activity is discussed in the next section. This represents the meaning of creation, which is distinct from the limited method of conventional design. To resolve the conflicting outcomes of design, a renewed design theme (motive) should be generated as a social motive of design. Regarding the previous message, the following critical issues are discussed to explain social motives in this chapter.

(1) Sustainability
We should identify reasons for design creativity to match social motives. Sustainability is an example of a social motive that is linked to the sense of nature. In this chapter, we discuss the reasons why design changes nature in conventional society while confirming a strong motive for sustainability. Based on this motive, there has been a strong emotionally driven tendency towards sustainable design since the 1990s (McLennan 2004; Chapman 2009).

(2) The inherent creativity of humans
The internal motivation of a person's inherent creativity and social motives must be bridged to ensure advanced design. The ideal relationship that develops between a designer and artifacts is surveyed using a "poietics" framework (Nagai et al. 2010). This can be expanded in "autopoietic" advanced design (self-organization process) frameworks as social systems in advanced society (Maturana and Varela 1980). In contrast, social motives were hidden under industry in conventional design. Hence, instead of reviewing conservative paintings using "common sense," the third section of this chapter will review abstract paintings (from various points throughout the history of fine art) from a cultural perspective.

(3) Incomplete design knowledge
We should understand that our knowledge of conventional design is problematically incomplete, especially from an ethics viewpoint. We also need to change the meaning of advanced society from a design creativity perspective because, as a force that can change or create design images (e.g., concept generation), this is an underlying cognitive process in design (Taura and Nagai 2012). These images motivate us to design and create products on both personal and social levels. Furthermore, we should grow our minds (inner criteria) toward future design, which will allow counter-production and create a safer future. For this, the actions of academics are expected to encourage future-oriented social motives. As such, this chapter provides a draft plan for academia and poses open questions.

The discussions of (1), (2), and (3) elucidate the meaning of design creativity in advanced society from a design motive perspective.

4.2 Sense of Nature

Nature is dynamic. That said, some changes are synthetic. In this case, the changed situation is a result of design. However, conventional design processes do not address the critical issue of why humans want to change the world. This section provides an overview of the current environmental situation caused by conventional design. We discuss social changes to realize motives such as sustainability.

4.2.1 Design Based on the Sense of Nature

Historically, humans have feared nature's uncontrolled power and might, which at times wrought immense damage on humans. However, man also benefited from nature. Patterns and periodicity exist in the natural world; thus, humans have sought to extract useful knowledge from it. The engineering design of today has also benefitted from understanding nature, and as such, this knowledge has formed the core principles of engineering. However, our principle knowledge of design is not completely extracted from nature. Here, design principles that protect the global environment are lacking.

The meaning of nature has shifted from something wild (barbarous, rough, simple, etc.) to something comfortable, useful, healthy, and valuable. Landscape paintings, healthy food, green gardens, wind power generation, and wooden kitchens are examples of this paradigm shift. These artifacts are valued according to their degree of "naturalness," and while they represent natural concepts, they are, interestingly, created by human technology. Therefore, humans can capture a sense of nature not only from natural objects, but also from manmade artifacts. The implication is that we can design artifacts that resonate with an embedded sense of nature. Conventional design hints at a sense of nature, but advanced design offers the potential to identify the essence of nature to create new designs that resonate with humans' inner sense. The idea that "design creativity" is essentially empowered by abstraction at a conceptual level is assumed. Indeed, fine art developed this type of creation in the twentieth century, for example, through abstract paintings that do not represent exact or actual sites of the natural world (Barr 1975; Dutton 1979; van Vliet 2009), but impact art history by expressing an embedded sense of nature. The basic motivation for art expression is fear of nature and hope for peaceful natural wellness.

This section focuses on the Pre-Design phase underlying social motives. This phase is strongly connected to motivation in cognitive processes and to reason in society. We can identify an inherent reason for design in humans—namely, the desire to survive and to expand into endless desire: in other words, a desire for a happier, wealthier life. During this expansion of desire, design motives have been identified as part of a general motivation for change to achieve an improved situation. Thus, in the conventional design framework, a design problem represents an undesired situation that must be solved by physical improvement (Rosenman et al. 1990). The seriousness of the problem—survival, convenience, or fame—is not usually discussed. Once the problem is determined, the goal of the design process is to immediately target it as an issue to be solved.

Usually, "changing the world" is interpreted in a positive way. However, we should recognize the negative side of this aspiration as well. For a long time, humans have designed the world and altered nature. We are afraid of the results of the long-term design activities of an industry-oriented society. Three hundred years ago, the social motives for design were primarily to protect humans from nature. However, in contrast, the social motives of today are to protect nature from humans. We can identify these strong environmental motives as tendencies toward sustainable design. A question that has yet to be addressed is whether we can apply our design knowledge (obtained in the past century) to design a future society that protects nature.

4.2.2 *Social Motives for Design in Nature*

"Sustainability" is an example of a social motive derived from a design perspective based on critical issues related to design and nature. We can follow the process of social events and design discussions related to this issue. The concept of "sustainability" is the clearest sign of an advanced society; in the past, this was an intangible concept. The critical issue is how the shift from intangible to tangible occurred. Additionally, academic contributions to this shift must also be confirmed, as they represent "social motives" and have the power to change design directions.

Young people are largely responsible for the strong movement toward social motives that privilege sustainability. "I am fighting for my future." Severn Suzuki's unforgettable speech at the United Nations Earth Summit in Rio de Janeiro stirred the emotions of people from many countries in 1992. Nobody could respond to her simple, but difficult, request: "If you don't know how to fix it, please stop breaking it." She concluded that "parents should be able to comfort their children by saying that 'everything's going to be all right. It's not the end of the world. And we're doing the best we can.' But I don't think you can say that to us anymore."

Twenty years later, we are reminded of her request while we reconsider the actions made to address her call.

To respond to her call, a future-orientated view of design—one that considers our children's future—is needed. To understand the embedded problems of our society and to become enlightened by overcoming the "ego," each individual must engage in careful self-investigation and become a deep thinker. Such awareness underlies design thinking, "deepening" it so that it can contribute to an advanced society.

Extinction is the sacrifice of design, that is to say, the result of human creativity. "Lonesome George," who died June 24, 2012, was the last Pinta Island tortoise ("Geochelone nigra abingdoni," a type of Galapagos tortoise). Certainly, the tortoise did not need the name bestowed upon him by humans as much as he needed the company of another tortoise of the same species. The WWF promotes the conservation of the natural environment and all animal species by using the living planet index (LPI), an indicator of the state of global biological diversity based on trends in vertebrate populations of species from around the world. Specifically, wild animals (e.g., tigers) have been suffering in recent years. In 2012, there were 28% fewer animals than recorded in 1970.

On the other hand, the press has reported new types of viruses, bacteria, and other microorganisms as the result of mutation: drug-resistant mutations like "NDM-1" (New Delhi metallo-beta-lactamase 1) exist because of these mutations.

This implies that humans are altering the natural system and unintentionally endangering lives: currently, no human on the planet can take responsibility for the results of design.

Extinct species are considered from a sustainable environment perspective. This aspect highlights the origin of human creativity as a selfish activity, driven by the "ego." Therefore, design progresses at a micro-level; macro-level design is lacking completely. While this unbalance was considered, halting the motivation for micro-level design is difficult. "Over the past 50 years, humans have changed ecosystems more rapidly and extensively than in any comparable period of time in human history, largely to meet rapidly growing demands for food, fresh water, timber, fiber, and fuel. This has resulted in a substantial and largely irreversible loss in the diversity of life on Earth" (Millennium Ecosystem Assessment 2005). Global warming will endanger species even further, particularly various plant species. Essentially, we could not change the world after 1992 without risking serious deterioration. This critical issue must be criticized among academics.

Can we conclude an unavoidable relationship between design and nature? We should replace older products as soon as possible if their design did not consider sustainability based on the social motive stemming from a sense of nature—the desire for design that works toward a sustainable earth.

4.3 Sense of Culture

4.3.1 Myth of Design

In the previous section, the social motive for design from a sense of nature was explained. These motives resonate with personal motivations and influence culture. Basically, design products inherit human emotions such as their wish to survive. Over time, human culture has been formed by a large number of artifacts that have internalized various human emotions. Recently, emotional design has become popular for product design arrangements; however, this is similar to an imitation, which lacks a deep emotional sense. Culture is a type of design technology that solves the problems of our current society. "Löwenfrau" is a famous statue created 32,000 years ago: the exact interpretation of this statue has not been identified yet. That said, Löwenfrau seems different from other cultural artifacts from the same time period. Löwenfrau's ornamental arrangement of shapes copied from wild animals, such as lions, tigers, and bulls, is probably more cognitively and culturally meaningful than strictly functional artifacts.

However, to elucidate the creative nature of human beings, there is no subject more interesting than design creativity. The "myth of design" refers to an unexplored region of science and engineering. In conventional design frameworks, one aspect of uncertain issues in the myth of design has been replaced by the expression "design is a wicked problem." In design, a more creative approach is needed; we need to, for example, create items to solve problems instead of continuing to use traditional approaches that promote determining the best solution from all feasible solutions.

From a view of design as embedded knowledge in society, this section explores the myth of design by discussing art design in connection with a sense of culture. The cognitive features of humans dynamically correspond to the constructed meaning of art. Thus, we can address the construction of meaning from a social cognition perspective. This process is characterized by an autonomous system (which forms a self-organizing system).

Usually, artifacts created through functional reason are categorized as crafts. Handcraft production processes hold the power of absorption. Such absorption is expressed as "Flow in Work," which is connected to a person's happiness beyond boredom and anxiety (Csikszentmihalyi 1996, 2003).

In conventional design, the notion of the "problem of design" was expressed to solve the sub-problems of society. Engineers or designers do not necessarily realize that their individual creativity is oriented toward the embedded knowledge of society. They have contributed to humanity's past prosperity without recognizing how it has disrupted the earth's balance. In the past, the original purpose of design was to beat competition in the market. Today, we are still trapped in this mind-set. Our generation should provide helpful sources to develop a new design framework for new design thinkers, even though we are not able to change the basic nature of human creativity that constructed society's "embedded motives." Usually, society's embedded knowledge generates design problems. The desire for a "better life" is an example. When the knowledge of humans dictated that the power of nature should be obeyed, we aimed for a simple life. Engineers and designers solved design problems to improve the usability and function of basic products. Indeed, these improved products reduce workloads related to housekeeping, working, manufacturing, etc. For example, brooms were replaced by an electric vacuum as recently as 20 years ago while even more recent developments have led to vacuums that automatically vacuum the floor. Moreover, several products were designed to enhance our lifestyles. Newly designed products provided fun and joy, making people's lives happier. These outcomes resulted from human knowledge. Therefore, we can represent embedded knowledge in society as the power to solve "embedded problems" for design, which forms a sense of culture.

4.3.2 Culture as a Mirror of the Social Motive of Design

What does it mean to speak of a sense of culture? This section focuses on the inner sense of individuals and the embedded sense of society from a cultural perspective. The former is beyond perception or cognition; rather, it is the natural appreciation or comprehension of the external world. This inner sense activates intuition or insight as an underlying motivation deep within the mind.

Fine art is an ideal example of a type of artifact that expresses social motives that form culture. Although many people confuse them, culture as embedded cognition of intangible social knowledge is different from common sense based on past customs. The power of the former resonates as a deep part of the human mind, namely, “deep emotion.” To clarify the embedded knowledge that enables design to drive the creative process, this section discusses creative fine art by focusing on the design process.

Fine art, especially painting, illustrates a long history of both concrete and abstract image expression. Famous creative artist Pablo Picasso wanted to break free from the traditional Western style of art. This pioneer of modern art is known for his many dramatic comments, some of which are thought to have become exaggerated as they spread through society. However, some of his remarks carry a deep meaning in regard to the creativity involved in art: “It took 4 years to paint like ‘Raffaello’ when I was a child. Moreover, to paint like a child needed more years, maybe a whole lifetime.” While this comment is cited in early childhood education literature, it suggests the limitations of common sense. Painting is an easy task that anyone can engage in: most children draw pictures before they write letters. The popularity of art forms the common sensual, societal notions of art: people possess the basic knowledge required to identify whether paintings are created by children, amateurs, or accomplished artists. This type of societal knowledge about art sets unsaid, high standards for artists. To become an artist, most painters study how to design paintings so that they meet social requirements. Basic knowledge about painting in regard to techniques, history, materials, and presentation can be obtained through education: however, the sense of culture (of art) must develop through each person’s experience. An inevitable challenge for the creative artist is to break away from such common sense and go against social requirements. Picasso’s aforementioned remark noted this issue. To paint like a great artist in the past is a silent requirement. However, the heart’s voice is stronger in more “childlike” expressions of art. A higher order of art is needed for originality. Poietics (originally “la poïétique”), proposed by Passeron, is the study of creative behavior. Poietics scholars maintain that creative activity is inherent to the motor nervous system, while aesthetics lie outside knowledge. Passeron denotes the creative activity of painting (for example) as an expression of specific sensitive phenomena. Essentially, poietics is separate from aesthetics; it comprises an autonomous system of internal knowledge, whereas the basis of conventional art design is the study of aesthetics and art history. Other conventional art design is aimed at answering the customers’ or audiences’ needs. This creative phenomenon reveals a structure that is similar to the circular interactive process of “doing–knowing.” Figure 4.2 demonstrates the basic framework of a self-organizing system of art creation in abstract paintings that is similar to the interactive “doing–knowing” process demonstrated in abstract art.

A self-organizing process in the autonomous system is obtained through a sense of culture. In this section, we focus on the social motive of “openness,” which occurs in the self-organic process of the autonomous system as a power separate from the (often closed) cycle of the creative evolution of originality.

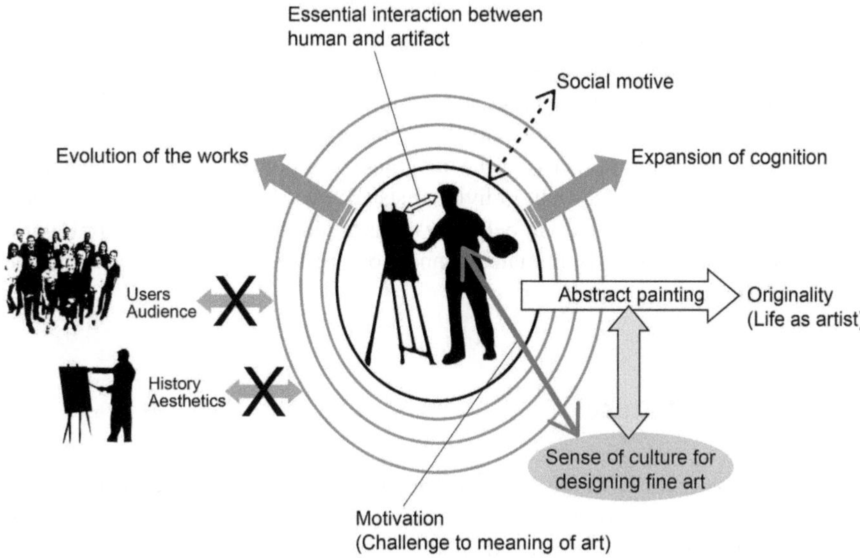

Fig. 4.2 Self-organizing system in creativity (Poietics)

Indeed, artisans who became absorbed in making artifacts seem happy with their elaborate work; however, the development of style is very limited. This is because the main purpose of the work is to repeat a traditional model. In other words, the problems are rooted in a strong feeling of attachment that limits the freedom of human creativity.

The radical artists of the last century, namely, the avant-garde, changed the tendency of repetition. They presented challenging paintings that differed from traditional styles. Kandinsky, Mondrian, Picasso, and Picabia are pioneers of abstract painting who express the essential images of the concept. All their works defy the traditional grammar of paintings; however, their paintings also represent the sense of culture reflected in social change. Although, these paintings are not traditional portraits of people, they still mirror peoples' motives in making an epoch of new citizens. Those motives were a desire for liberation from systematized society and were achieved by Stella, Judd, LeWitt, and so on. Moreover, the movement of abstract painting set off a chain reaction in various parts of the world. Both Pollock and Twombly presented unique paintings that integrated a self-organic process similar to the "poietics" and avant-garde styles, which worked against the common sense of fine art. Their works were deeply personal, but they still resonated with many other people. This example allows us to understand the essential function of the sense of culture that generates the next social motive.

As we mentioned in the previous section, people design their surroundings to enhance their environment. Unfortunately, this enhancing of human environments often comes into conflict with, and weakens, the earth's environment. Such problems, inherent in our society, must be examined from a design perspective. The origin of these "closed system" problems is probably the popular enthusiasm for making objects, including those that are designed (Csikszentmihalyi and Rochberg-Halton 1981). Although these problems are complex and intertwined, we attempt to classify them to find a suitable approach to identify the sense of culture. The social brain hypothesis (e.g., the "Machiavellian intelligence" hypothesis) models the human brain's evolution process as the coevolution of humans, who are fundamentally a social species rather than individualists (Byrne and Whiten 1989). Accordingly, scholars claim that human behavior tends to aim to govern the natural world by forming culture. This illuminates connections between social motives and the senses of nature and culture.

4.4 Sense of Future

In the previous section, we discussed the interactive relationship between humans and artifacts as a creative activity. The main goal of most design initiatives is to produce products; the unintentional final outcome of products is the state of the world's environment. This is the result of a closed system. In the cycles of the creative process, human cognition seems to be separate from outside factors. To break free from the closed system of conventional design, the sense of future must be discussed. In doing so, we hope to increase awareness of a sense of the future.

From an ethics perspective, the reasons for creativity are essentially different from the reasons for social design. Notably, neuroethics (biomedical ethics) has focused on detecting brain functions to answer the question of why only humans can have a motivation for the actions of others (Farah 2010; Bruni et al. 2013). These features connect to deep design thinking in future-minded individuals.

4.4.1 Deep Design Thinkers

In the twentieth century, problems related to the embodied incoherence of industrial society tended to be uncontrolled. Correspondingly, aspects of design studies, research, discussions, and education have grown rapidly without much criticism. Before completing theoretical academic studies of design, we must develop design students to be human resources who are able to contribute to an industrial society through their design ability. Design education models that privilege realistic design practice—such as experiential learning that is connected to the real world—over philosophical reflection (however important and relevant) are needed.

This chapter identifies the roles of academics in regard to designing an advanced society, which should differ from academic roles in conventional society. The history of the nineteenth and twentieth centuries developed a firm foundation for production and manufacturing, which is usually understood as a restricted meaning of design. This restricted meaning of design fits the framework of conventional design. Much of the development of science and technology is driven by academics. Academia requires scholars to explore the behavior of nature and attempt to understand its mechanisms and functions. This made sense in industrial society, and the basic ideas of conventional design were reflected in evaluation criteria that required products to be "useful," "rational," and "economical." Most design researchers have contributed to generating and adding beneficial knowledge to increase the usefulness, rationality, and economic value of products. Accordingly, the aims of university education have tended to cultivate students who will contribute to industry and society after graduation. In the twenty-first century, this tendency has become increasingly important. A number of studies on design highlighted the importance of "originality" in both product and engineering design. However, in contrast to art or science, originality has not been scrutinized in conventional design. Rather, the "originality" of a design was included as an "economical" criterion in industrial society. It is supposed that imitating a product becomes common sense when it is not protected by patent and copyright systems. Imitation can be part of the production system if expenses are paid for permission in conventional design. Naturally, in art and science, originality is valued. That said, it is not to prevent imitation, but to protect the authority and reputation of the author. Indeed, many imitations sometimes raise the value of the original work.

On the other hand, in advanced society, an advanced design framework is needed. To reform a framework of design from a design creativity perspective, embedded concepts in society, nature, culture, and future will be incorporated as the internal components of the system.

Papanek's critical view sheds light on this issue (Papanek 1971, 1995). In the 1970s and 1980s, he claimed that one must consider "design for humans"—effectively suggesting that design for customers in the 1990s must explore ways to live in harmony, instead of in conflict, with nature. His critical essay about industrial culture and a society that represents mass production is especially valuable. Among design students in the late twentieth century, his message that "design has become the most powerful tool with which man shapes his tools and environments" has been discussed from two opposing perspectives—namely, a positive and negative intention. Furthermore, future subjects are pointed out from a design perspective for social innovation. This aspect highlights expectations for new "design thinkers." Educating design thinkers has recently become more popular; creating experiential challenges that lead students to identify societal problems tends to generate innovative ideas to change the world as well as future prospects.

Presently, design thinkers are expected to be educated leaders who can identify societal problems, help design society, and explore the future world. Design thinkers can be the new model of creativity at the frontier of the next era. Images of design thinkers differ to those of stereotyped engineers and scientists. Design thinkers are seen as business risk-takers with rich social skills that enable them to forge collaborations among other people. "Design creativity" is a concept representing the integration of a creative mind, creative knowledge, creative technology, and a suitable attitude, which will be gained by design thinkers. "Design creativity" is necessary if we are to go beyond past designs and educate ourselves to solve societal problems created by humanity.

To develop such "design creativity," it is important to distinguish the essential issues and establish a suitable methodology to cultivate design thinking. Multiple views are required to process design thinking; however, education programs might misunderstand what is meant by "interdisciplinary." Neither T-type nor π-type integration between multiple disciplines is enough to foster "a deep thinker." This discussion indicates the importance of fostering "deep design thinkers" who are socially motivated and "committed" to drive advanced society into the future. These notions of commitment represent the soul of socially minded participants.

4.4.2 Actions Toward the Future

Will we not regret our actions if we leave a human-made world of difficult problems for our children? In engineering and architecture, "design" in a physical regard is defined as something that changes a problematic situation into a preferred one. So far, how have recent engineering and other design domains contributed to averting the difficulties our children face?

The last part of this chapter discusses the meaningful roles academics can play to create a more sustainable and ideal advanced society in regard to not only the pragmatic, but also the ethical, meaning of the Pre-Design, Design, and Post-Design phases. To encourage academia to deepen its commitment to society, a draft action plan based on "design creativity" that contributes to our future society is considered in the last section. In the future, an ideal design community will be formed by academia and society; however, many uncertainties remain that should be clarified to ensure a link between academia and society. For example, the responsibility for a design discourse is a complicated problem.

The challenge of connecting science and ethics is becoming a more common issue in academia, especially in the cognitive science field of the human and social sciences (Johnson 2010). Profound knowledge of total design, as well as the courageous action of academics, is required to balance science and ethics in design.

Currently, academics play a more important role in defining programs that develop design thinkers than in educating good engineers, technologists, and scientists who are already industry-ready. To play that role and answer the aforementioned call made by a girl representing a children's eco group 20 years ago, we must first carefully check the results of previous designs driven by our society. Second, we need to consider if lost human technology and knowledge holds clues that can solve humanity's embedded problems that have changed nature, formed culture, and designed our present society.

This action plan will connect business opportunities with present needs. However, other issues not provided by teams or users should also be considered. For example, animals, plant species, the knowledge of traditional architects, the intangible properties of handcrafts and art should be protected. Further, to develop a

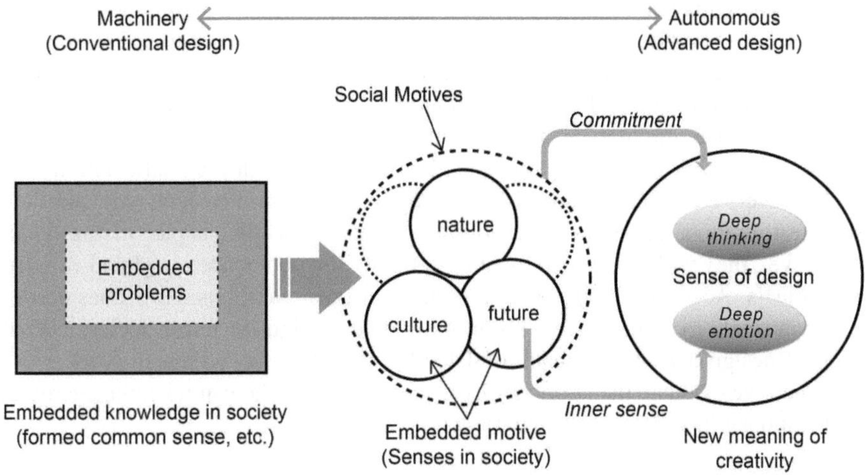

Fig. 4.3 Framework of a new meaning of creativity in advanced society

truly interdisciplinary design program that cultivates deep thinkers, there must be a precise process that enhances creativity.

Taking into account social motives from nature, culture, and future perspectives, this chapter identified the growth of a sense of design as the importance of design creativity in an advanced society. "The meaning of the growth of a sense of design" has generated a new social image of the future that resonates with deep human emotion among profound design thinkers. Creativity in advanced design is characterized by a purely self-organizing system that is internally defined and interactively influenced from the outside (from the Post-Design phase); it generates images that resonate with the core senses of design—nature, culture, and the future—embedded in society (see Fig. 4.3). A rare example of this cycle is found in designing fine art through the exploration of abstract painting because "abstract" and "realistic" representations are often in sharp conflict with one another. Outstanding artists challenged traditional painting concepts by exploring the sense of culture and, as a result, mediated a social motive of "autonomy" in abstract art as a self-organizing system. However, in design, we found that the sense of caring for life, which resonates with people's deep emotions, has triggered environmental social motives, which have led to the new challenge of designing sustainable products, services, and systems. This will be the next critical issue for deep design thinkers as they become future-oriented creators who go beyond the selfish requirements of users to upgrade the value of products, services, and systems. Therefore, commitment and ethics will become mandatory virtues of design creators.

To contribute to the above issue, academics must act from a place of heartfelt concern for the actual world. This means that academics should work toward fostering sustainable design and the next generation of serious thinkers, while encouraging a commitment to a future advanced community that drives creative design. There may be other ways to lead this movement, but a pragmatic attitude is needed. For example, each person must oblige himself or herself to spend of his or her time on these objectives. Based on the discussion of this chapter, drafting a globally collaborative and relevant action plan, namely, the "effort TEN project," that calls for the voluntary acts of academics is an idea that should be proposed to the design community to develop social knowledge through pragmatic activity.

However, the owner of social knowledge is uncertain. Problems will publicly manifest as moot questions (open questions) based on this chapter of the book. Moreover, they will be updated by other scholars and non-academic global-minded individuals. Accordingly, there are a number of open questions. How does academia contribute to culture? How might individuals design the future society? What kind of design activity is meaningful for future society according to academics? How can we evaluate the applicability of our actions? Who can take management, and responsibility, of such voluntary acts?

4.5 Conclusion

Advanced design was addressed on the basis of a Pre-Design, Design, and Post-Design model by using three aspects to connect each phase. A new meaning of creativity in advanced design was explained as a self-organizing system that can be defined from the inside, which generates images that resonate with the core senses of design—namely, nature, culture, and future—embedded in society. The social motives of advanced design that represent "sustainability" from a sense of nature, "openness" as a self-organizing system from a sense of culture, and "deep design thinking" from a sense of the future were extracted throughout the discussion of critical issues in regard to creativity in design in human and social cognition. In the future, studies concerning the connection of abstract images in art, and idea generation in design, will need an approach to understand why humans can generate and apply such abstract concepts to develop a design sense and extract the cognitive features of humans in activation processes.

A draft plan for design academics was proposed to progress design creativity to correspond to the extracted motives.

References

Barr AH (1975) Cubism and abstract art. Secker & Warburg, Buckinghamshire

Bruni T, Mameli M, Rini RA (2013) The science of morality and its normative implications. Neuroethics. doi:10.1007/s12152-013-9191-y

Byrne RW, Whiten A (1989) Machiavellian intelligence: social expertise and the evolution of intellect in monkeys, apes, and humans. Oxford Science Publications, Oxford University Press, Oxford

Chapman J (2009) Design for emotional durability. Design Issues 25(4):29–35

Csikszentmihalyi M (1996) Creativity: flow and the psychology of discovery and invention. Harper Perennial, New York

Csikszentmihalyi M (2003) Good business: leadership, flow, and the making of meaning. Penguin Group, New York

Csikszentmihalyi M, Rochberg-Halton E (1981) The meaning of things: domestic symbols and the self. Cambridge University Press, Cambridge

Dutton EP (1979) Abstract painting. Arsen Pohribny, Boston

Editorial Board of IJDCI (2013) Perspectives on design creativity and innovation research. Int J Des Creat Innov 1(1):1–42. doi:10.1080/21650349.2013.754657

Farah MJ (2010) Neuroethics: an introduction with readings (basic bioethics). MIT, Massachusetts

Johnson M (1987) The body in the mind: the bodily basis of meaning, imagination, and reason. University of Chicago Press, Chicago

Johnson M (2010) In: Ambrose D, Cross T (eds) What cognitive science brings to ethics in morality, ethics, and gifted minds. Springer, London, pp 147–150

Lakoff G (1987) Women, fire, and dangerous things: what categories reveal about the mind. University of Chicago Press, Chicago

Maturana H, Varela F (1980) Autopoiesis and cognition: the realization of the living. Boston Studies in the Philosophy of Science, p 42

McLennan JF (2004) The philosophy of sustainable design. Ecotone Publishing Company LLC, Kansas City

Millennium Ecosystem Assessment (2005) http://www.maweb.org/en/index.aspx

Nagai Y, Taura T, Sano K (2010) Research methodology for internal observation of design thinking in the creative self-formation process. The first international conference on design creativity (ICDC2010), pp 215–222

Papanek V (1971) Design for the real world: human ecology and social change. Pantheon Books, New York

Papanek V (1995) The green imperative: natural design for the real world. Thames and Hudson, New York

Rosenman MA, Radford AD, Balachandran M, Gero JS, Coyne R (1990) Knowledge-based design systems (the Teknowledge series in knowledge engineering). Addison-Wesley, Pearson, Boston

Taura T (2014) The design of technology: bridging highly advanced science & technology with society through the creation of products, in pre-design, design, and post-design: principia designae for the highly advanced technological society. Springer, Japan

Taura T, Nagai Y (2012) Concept generation for design creativity: a systematized theory and methodology. Springer, London

van Vliet R (2009) The art of abstract painting: a guide to creativity and free expression. Search Press Ltd., Turnbridge Wells

Chapter 5
The Pagoda Design Space: Extending the Scope of Design

Gabriela Goldschmidt

Abstract A comprehensive notion of design that includes pre-design and post-design is presented in this chapter as hinging on an extended design space. Not just the array of possible solutions, the design space is modeled as comprising two main components. The first is design expertise, which is composed of five tiers: pre-design, task framing, design acts, design proposal, and post-design. The second component is creativity, which is embodied in an insight-impact axis that cuts through the various expertise layers. The model is metaphorically likened to the indestructible Japanese five-tier pagoda, whose strength emanates from flexibility and independence of components, which interact but are also unattached to other components.

Keywords Design space • Expertise • Impact • Insight • Pagoda

The five-tier pagoda, a building type of which a few ancient specimens survive in Japan, is an impressive structure that has proven to be extraordinarily durable in the face of numerous harsh earthquakes that destroyed other contemporary buildings. A structural analysis attributes this durability to a high level of resilience that is gained by avoiding rigid connections within and between the different components. This allows maximum movement of each component relative to the other components.

I use the pagoda as a metaphor for a new, extended model of the design space. Traditionally the design space is seen as the aggregation of all possible design solutions in a given task. Here it is conceived as encompassing design knowledge, or expertise, that guides designing from pre-design to post-design, through framing, exploration, assessment and the choice and development of the most appropriate design solution. The layers of expertise are arranged around an independent central axis that leads from insight to impact, cutting through the body of expertise. This axis is reminiscent of the central pillar of the five-tier pagoda, which is not anchored to the ground and around which the five stories are built, unattached to it, but regulating its motion. This axis stands for creativity. I conclude that a model

G. Goldschmidt (✉)
Technion—Israel Institute of Technology, Haifa, Israel
e-mail: gabig@technion.ac.il

© Springer Japan 2015

T. Taura (ed.), *Principia Designae – Pre-Design, Design, and Post-Design*,
DOI 10.1007/978-4-431-54403-6_5

combining expertise and design creativity in one design space in which different components are independent yet free to interact in a complex network of links, is a more affording account of the notion of design that extends from pre-design to post-design.

5.1 Design Space

The term *design space* was coined in artificial intelligence (AI). It has a wide range of interpretations but typically it denotes the "space of possible designs for behaving systems" (Sloman 1995, p. 1), where the systems may be natural or artificial, and they include "architectures, mechanisms, formalisms, inference systems, and the like" (Sloman 1995). The design space is explored in order to "characterize diverse behavioral capabilities and the environments in which they are required, or possible" (Sloman 1995). Design space exploration (DSE) has become a major field of study in AI and engineering, especially software design. In engineering design, or problem-solving, "the design space is a representation of all possible solutions" (Westerlund 2005, p. 1), and design space exploration suggests exploring alternative solutions (e.g., Saxena and Karsai 2010; Woodbury and Burrow 2006). Instances of design, or alternatives, inhabit the design space in the form of representations. In engineering we encounter the model-driven design approach, wherein these representations are often models, and accordingly we find definitions such as: "Design space exploration (DSE) aims at searching through various models representing different design candidates" (Hegedüs et al. 2011, p. 1). The purpose of DSE is to arrive at a high quality solution in a cost-effective manner while ensuring that a sufficient number of alternatives are compared and assessed against goals and constraints. According to Woodbury and Burrow (2006), DSE research typically "addresses representation, search algorithms, task description, or interaction design… most research focuses on design states and on making action explicit" (p. 64). They go on to state that "relatively little work focuses on the design space itself… Yet, it would appear that the design space itself is where the largest gains are to be made" (Woodbury and Burrow 2006). As this research topic is based in AI, most of the work on the design space concerns itself with the computability of design spaces and computational access to them (Woodbury and Burrow 2006). However, the design space, albeit under different appellations, has also been addressed by researchers from other fields, notably architecture. Some of this work is quite old by now, and does not refer to computation. We shall briefly review samples of this work, which is relevant to the current discussion.

When the notions of problem space and later solution space were introduced (Newell and Simon 1972), models of problem solving in various domains were crafted to fit the principles on which they were based. Beside a goal, the basic problem space contained a set of states of knowledge, and operators by which states are changed from one to another. In addition, the problem space encompassed constraints on applying operators, and what was termed 'control knowledge' by

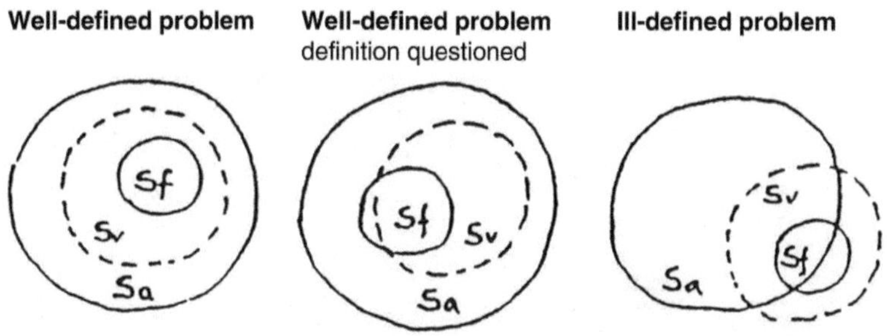

Fig. 5.1 Solution spaces for differently defined problems (Habraken 1985, p. 68). Solution spaces: *Sa* within explicit problem constraints; *Sv* virtual – valid for problem solver; *Sf* forms acceptable as candidate solutions

which the problem solver decided which operator to apply next. Today operators are algorithms and they include the constraints. In software design the notions of problem and solution spaces is still in wide use, and mapping from the problem space to the solution space is a widely researched topic.

In architectural design the concept of problem space received a wider interpretation and was closely related to the search aspect of the space, sometimes also termed search space. The concern in architecture was not just efficient problem solving, but coming up with a creative solution, or design. As such, the issue of the space's boundary gained importance. Likewise, it was acknowledged that a problem is often too complex to solve without breaking it up into partial sub-problems. Unlike the AI notions, the terms problem space and solution space were used interchangeably in the early design literature, as we shall see in the examples that follow. At a later stage it was suggested that the two spaces should be conceptualized as a single space within which design problems are searched and solutions are developed (Dorst and Cross 2001; Maher et al. 1996). Goldschmidt (1997) wrote about a design problem space, but an extended one in which a solution is in fact sought. I equate this space with the design space.

Let us consider a couple of examples from the design literature that were concerned with creative design. The first example is from a text by John Habraken (1985). Habraken distinguished between well and ill-defined problems and established three types of solution spaces. The widest one Sa, is a space containing all possible solutions under explicit problem constraints. The second, smaller space Sv, is virtual and limits the solutions to those valid for the problem-solver. The last and smallest space Sf, includes only forms that are acceptable as candidate solutions. Figure 5.1 depicts three arrangements of solution spaces, varying according to the level of definition of the problem.

We note that in the case of well-defined problems the three spaces are concentric and well aligned one within the other, with no violation of space boundaries. As the problem becomes less defined, that is, the definition is questioned, the space of acceptable forms exceeds the realm of the designer's initial valid solutions, as questioning gives rise to new possibilities. Finally, when the problem is ill-defined

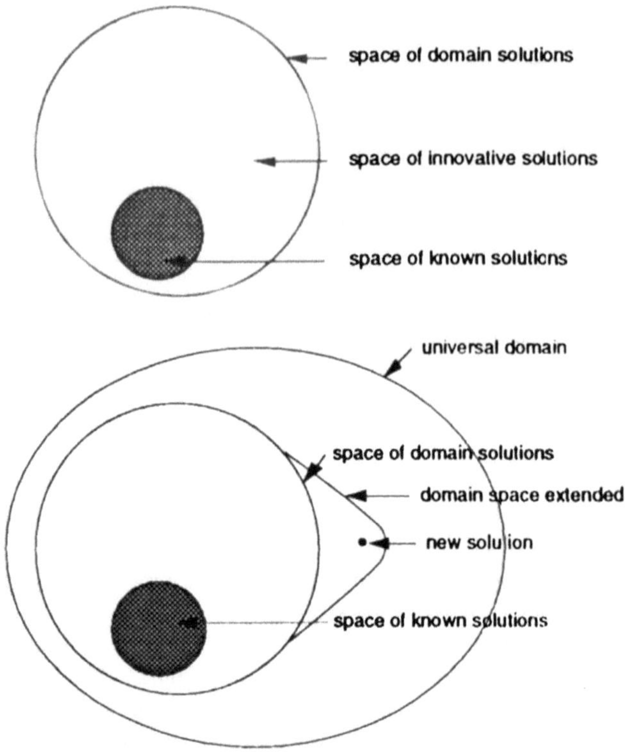

Fig. 5.2 Solution spaces. *Top*: Innovative design. *Bottom*: Creative Design (Rosenman and Gero 1993, pp. 114–115)

and the designer embarks on an extensive search, new knowledge is gained which yields solutions that are not only beyond the designer's prior concepts, but also beyond what had been considered acceptable within the constraints. In other words, the boundaries can be pushed if the designer does not stick to known and acceptable solutions but charts new territory.

A similar idea was expressed by Rosenman and Gero (1993) who associated an extended solution space with creativity. Rosenman and Gero distinguished among routine design, which involves the adaptation of a known solution; innovative design, "in which the space of known solutions is extended by making variations or adaptations to existing design" (Rosenman and Gero 1993, p. 113), and creative design, which "involves the generation of entirely new types [of solutions]" (Rosenman and Gero 1993, p. 114). In the latter case the boundaries of the solution space are pushed not only beyond the space of known solutions, but also beyond the space of innovative solutions. Figure 5.2 is an illustration of an innovative and a creative solution space as proposed by Rosenman and Gero.

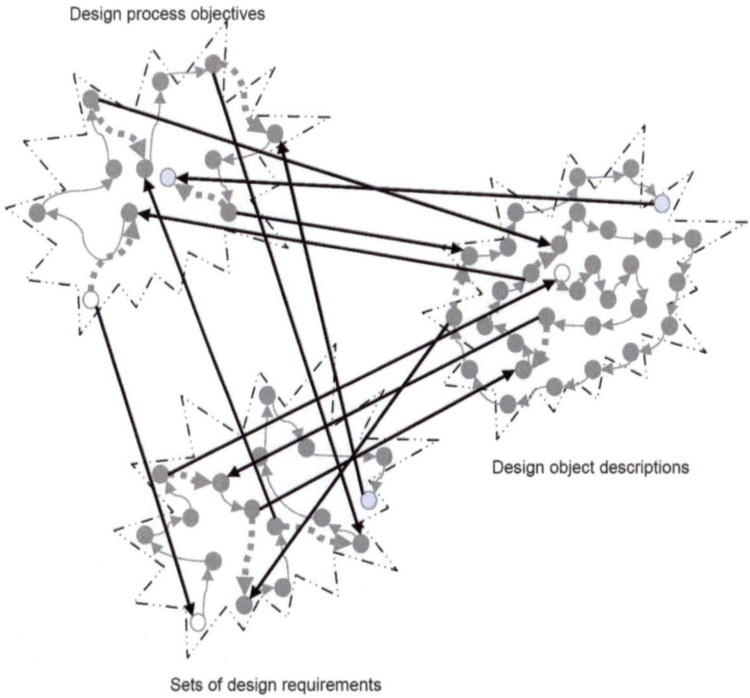

Fig. 5.3 Exploration links within and between design subspaces (van Langen and Brazier 2006, p. 116)

Contemporary AI researchers take a somewhat different approach. In their view the boundary of the design space, which integrates problem and solution spaces, are flexible a priori and define the realm of undiscovered designs, or possible solutions. Because of the complexity of design problems some researchers suggest separate, yet interconnected, subspaces in which different design knowledge is reasoned about. For example, van Langen and Brazier (2006) present three such spaces, concerning design process objectives, design object descriptions, and sets of design requirements (Fig. 5.3). There are links among objects within each subspace, and links among objects in different subspaces. The links represent acts of exploration.

Woodbury and Burrow's (2006) notion of the design space focuses on the relationship between discovered and undiscovered designs, where 'design' is an instance of design, or a designed entity. Within the flexible boundaries of the undiscovered design space we find enclosed discovered subspaces of known design instances. This is shown in Fig. 5.4. The links, or paths, between instances of discovered and undiscovered designs are called tendrils and they stand for design exploration. The shorter the tendril, the less effort is required to arrive at the undiscovered design instance. Undiscovered designs that have paths, or connectivity, to a large number of discovered designs—are advantageous.

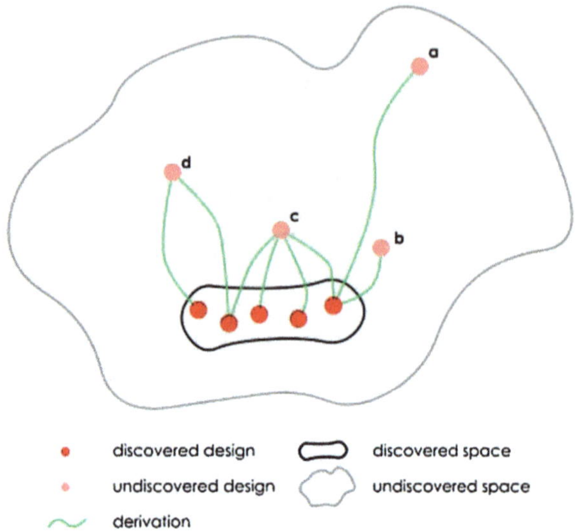

Fig. 5.4 Links among discovered and undiscovered designs in the design space (Woodbury and Burrow 2006, p. 66)

While the early work by Habraken (1985) and Rosenman and Gero (1993) stresses the boundary of the design space, the later work by van Langen and Brazier (2006) and Woodbury and Burrow (2006) stresses linkage in design exploration. The motivation of the former work is the study of design creativity, while the latter works are driven by information processing and the study of arrival at satisficing solutions. In both cases the design space is populated only with instances of design (with the possible exception of van Langen and Brazier, who hint at process objectives as part of the design space); they do recognize issues beyond the immediate search concerning design solutions. We shall now move to a brief discussion regarding the need to expand the view of the design space, to include pre and post-design.

5.2 The Need to Widen the Scope of Design

Donald Schön coined the term *design world* (e.g., Schön 1983, 1991). He stated that he used this term in the spirit of Nelson Goodman's (1978) *Ways of worldmaking*, that is, holding a constructionist view, which maintains that "our perceptions, appreciations, and beliefs are rooted in worlds of our own making that we come to accept as reality" (Schön 1987, p. 36). Schön thus proposed that the designer interacts with a design situation which is for the most part material, and which is appreciated by the designer through action, in part sensory action such as drawing. Through the apprehension and appreciation of the situation the designer "constructs and reconstructs the objects and relations with which he deals… thereby creating a "design world" within which he functions." Schön goes on to assert that "a design world may be unique to a designer or may be shared with a larger design community… Designing is primarily social…" (Schön 1991, p. 4). If we map the concept of design world onto the design space, which can be done albeit differences between the two, we interpret Schön as insisting that something that is common to a community (professional or otherwise) is actually part of the design space in which the designer operates. I take this to mean that even before a design task is formulated, there are already a host of factors that shape the design situation a priori. These include what is known about social and cultural conventions and traditions, the history of a place or building type or type of product, the technological state of the art, environmental issues, legal concerns, prevailing design methodologies, and more. But pre-design also includes the designer or designers' beliefs and values, both in general and as regards the professional domain to which they belong (for example, adherence to a particular style), various constraints, and goals beyond those stated in the task brief (for example, an architect may want a building to completely merge with nature, although no such requirement was advanced). In other words, the designer does not start from scratch: he or she or they start at a spot which has clear coordinates that must be addressed.

Pre-design knowledge is the platform on which a design exploration begins, and during the exploration things move around in the design space as priorities shift, information is interpreted from different points of view, relationships are defined and redefined, and a succession of mental models, individual as well as shared, are created, revised, recreated, and so on. This is the process of framing the design problem, task or situation. Following the initial framing (as well as in parallel to it), an exploration gets under way which consists mainly of embodying the entity that is being designed (or parts thereof) and establishing a rationale for embodiment acts (Goldschmidt 2012). Concrete propositions, partial and complete, are made and assessed and passed on to a more articulate phase of detailing. Upon the emergence of the most appropriate response(s) to the initial situation or one that arose from its manipulation, a design proposal is developed. The proposal or proposals are then typically presented to the stakeholders in pursue of their approval. If the project moves ahead, at that point the design process shifts to construction or production supervision, until it is brought to completion. Normally this is considered the end of the design cycle.

So far we have seen that pre-design is tightly connected to design, where the latter is the nominal activity of problem solving after a design assignment is launched. Can we be satisfied terminating this cycle when the artifact is materialized, that is, the product manufactured or assembled, or the building constructed? The answer is clearly negative. A need to gather feedback from the designed entity after it is put to use was evident to methodologists already in the era of the 'design methods movements' in the 1960s, mostly in architecture. The idea was that buildings do not always perform as expected once they are occupied and from a user-centered point of view it is important to learn about gaps in performance and use. The lessons learned are implemented in preparing for new designs, especially of same building types (e.g., health facilities, jails). The name given to the systematic gathering of feedback was Post Occupancy Evaluation (POE). A useful definition states that "POE is the examination of the effectiveness for human users of occupied designed environments" (Zimring and Reizenstein 1980). The term Post Implementation Evaluation can also be found in the literature. By extension, I propose the term Post Use Evaluation (PUE) that refers to feedback regarding the use of products.

POE continues to be in wide use (see e.g., Wall and Shea 2013), which attests to a comprehension that the design lifecycle does not terminate with the materialization of an artifact, but rather that there is an aftermath. In the case of POE, PUE, and the like, this aftermath is rather practical: the feedback is used to provide better guidelines for future designs of artifacts in the same category. I would like to claim that beyond immediate practical implications we should be interested in a post-design phase because this is when the effects of the design are revealed. Designers know that a good design should benefit society at large, and not only the direct users or sponsors of a particular artifact. In extreme cases the impact is far reaching and the design serves as a prototype or inspiration for many future designs, independent of the degree to which it performs its intended function. There are many examples of designed artifacts that have had a profound cultural impact long after the product was put in the market or erected. Apple's iphone is an obvious example, as it gave rise to the entire industry of smartphones. The Sydney Opera House is a building that although its design had won a competition, was constructed with great difficulty and at a cost that grossly exceeded the budget, and therefore its architect was fired in the process. Nevertheless the building, which suffers from functional shortcomings (e.g., seats with a partial view of the stage in a performance hall) has become an iconic symbol of Sydney. In a belated act of recognition the architect, Jørn Utzon, who was for a long time a persona non grata in Australia because of the problems associated with the building, was invited to return to Sydney as a guest of honor. Naturally most buildings and artifacts do not have a dramatic social and cultural impact that can compare to the iphone or the Sydney Opera, but they certainly can make a change in the lives of individuals and communities and they are always sources for feedback that can improve further design.

We see that the lifecycle of a design can and should encompass a pre-design phase and a post-design chapter to complete the picture, as the pre- and post-design portions are those that anchor the design in the culture in which it is rooted and on which it has an impact, large or small.

5.3 The Pagoda Structure

To explicate the design process, including the pre-design and post-design phases, I would like to use a metaphor, that of the five-tier pagoda. The metaphor will be described in the next section; in this section the pagoda structure is briefly introduced.

The five-tier pagoda: Gojū-no-tō (see Fig. 5.5) in Nara, Japan, which is part of the compound of the Buddhist temple *Kōfuku-ji*, was built in 1426 and is one of a

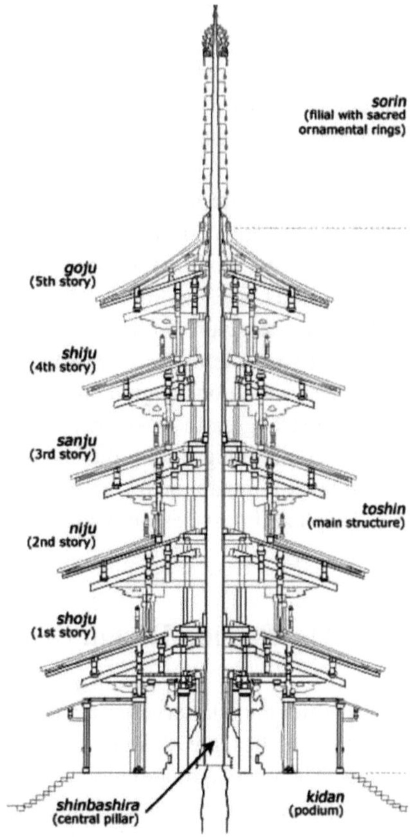

Fig. 5.5 Section through a five-tier pagoda. http://web-japan.org/nipponia/nipponia33/en/topic/index02.html

small number of such wooden structures to survive in Japan (very few additional examples can be found in China and Korea). The survival of pagodas of this type (there are three-tier pagodas of similar design), while so many other contemporary and later structures were destroyed by earthquakes over the years, is wondrous. It begs a look into the structural properties that made them withstand the harsh jolts they had suffered more than once. Figure 5.5 shows the structure to consist of five stories that are arranged around a central pillar, *shin-bashira*, which protrudes beyond the top of the fifth story to crown the pagoda with a symbolic ornamental top, *Sorin*.

Nakahara et al. (2000) explain that structurally, three features stand out in the pagoda: First, the central pillar does not have a foundation in the ground, as one would normally expect; it just rests on the base stone in the podium. Second, the five stories are also not attached to the ground or to each other; each story rests on the previous one and encircles the central pillar, without being connected to it. Third, each story is constructed of wooden members that are connected to one another with many complex joints. Japanese wooden construction is famous for its sophisticated dry joints, which depend on a perfect fit between members, without any nails or glue.

How do these features help the structure to resist destructive forces such as major earthquakes? The independence of the central pillar allows it to swing when subjected to extreme forces. The amplitude of the swing is constrained by the weight of the surrounding wooden stories, which swing along with it. Because the entire structure is not connected to the ground, no breakage is caused as free movement is enabled. The structural independence of the individual stories also contributes to a flexible motion of each vis a vis its neighbors, and since each story has a limited mass, the impact of the blow is much smaller than it would have been if the entire structure were a single, much heavier mass. Likewise, the many dry joints between members within the larger structure of each story, allow mutual movement which once again, at this smaller scale, restrains the impact even in the case of a strong earthquake. All these factors interact and together they safeguard the structural integrity of the pagoda. We should now like to use the pagoda structure as a metaphor for an extended model of the design space, including the pre-design and post-design phases.

5.4 The Pagoda Design Space: Creativity and Expertise

The five-tier pagoda serves as a metaphor for an extended model of the design space which embodies the design process, including the pre-design and post-design phases. The pagoda model has an independent central axis with five tiers surrounding it, that rest on each other. The central "pillar" extends from *insight* at its bottom to *impact* at the top, crowning the model as the *sorin* crowns the pagoda. The five "tiers" are the main components of the design process, namely, from bottom to top: pre-design, task framing, design acts, design proposal, and post-design. Figure 5.6 is a diagram of the pagoda design space model. In what follows, we briefly articulate the components of the main body of this design space and comment on its relationship with the central insight-impact axis.

5.4.1 Tier 1: Pre-design—State of Knowledge

Section 5.2 above explained in some detail what is included in pre-design and why it is part and parcel of the design world which is already in place prior to any specific activity of designing. Professional expertise consists of knowledge and know-how; knowledge is partly 'objective' and partly 'subjective'. The pagoda design space model spells out the main realms of "objective" knowledge: Historical, that is, architectural precedents and references; Social and cultural, pertaining to norms and conventions regarding the needs, expectations and lifestyle of particular populations. These norms may be dynamic of course (for example, a few

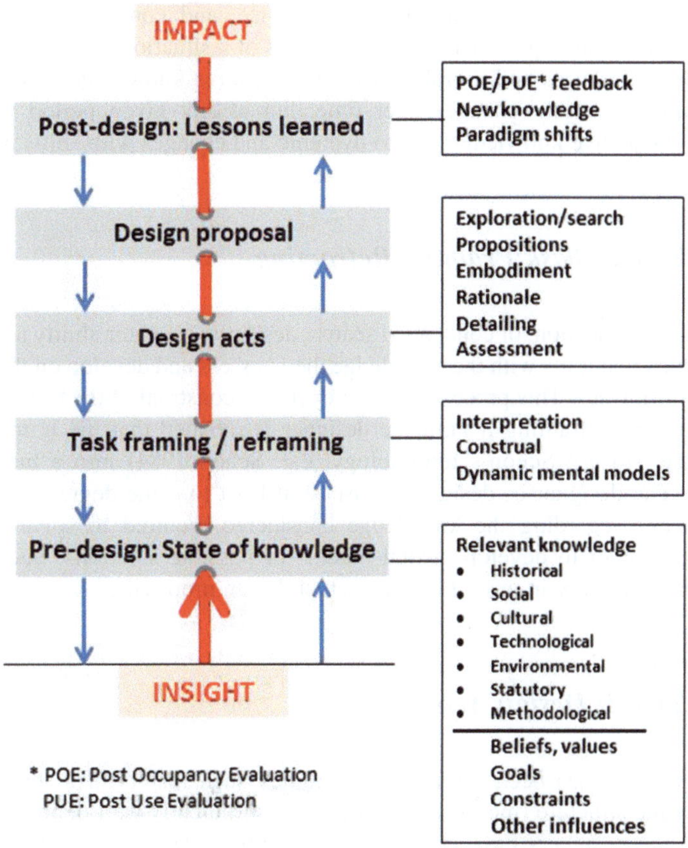

Fig. 5.6 The pagoda model of the design space

decades ago, projects of housing up to four stories high often did not include elevators; it was normal to walk up stairs to reach the higher floors. Today people are no longer willing to climb up a few floors and they expect to use elevators). Next is technological knowledge, which pertains to materials and to production (or construction) techniques, as well as to design support tools. Environmental knowledge caters to issues of sustainability at an ever growing level of sensitivity; Statutory knowledge, which designers must have in order to abide by safety and well-being regulations along with other legal requirements; and finally, methodological knowledge, which includes formal and informal design methodologies that are in common use in specific design disciplines.

In addition to these types of "objective" knowledge, the pre-design component of the design space includes factors that may be particular to an individual designer or to a group that embraces its own agenda. This knowledge includes beliefs and values which the designers hold and to which they subscribe, goals derived from a world-view, be it political, artistic, or otherwise, and constraints that do not emanate from legal requirements or the specifics of a situation but rather from the designer's preferences and moral codes. All subjective knowledge is dynamic of course, and frequently changes over time, but at any given period it can be specified. Objective knowledge is also dynamic and changes with shifts in norms.

5.4.2 Tier 2: Task Framing/Reframing

At the phase of ideation, or conceptual search, designers use their ability to interpret the task such that it fits with the knowledge they possess and develop on the fly, and with their priorities. This process calls for cycles of construal of the task that assist in framing and reframing it until the designer is satisfied that the frame turns a messy state (to use Schön's terminology, e.g. Schön 1984) into a manageable situation. The designer or designers must be at least to some degree within their comfort zone regarding the knowledge considered required to solve a design problem, in order to develop mental models of possible solutions. Such mental models are necessary in order to create actual design proposals.

5.4.3 Tier 3: Design Acts

Once a problem has been framed and reframed such that from a state of poor definition and structure (messy state) it becomes sufficiently well-defined, within a paradigm the designers have chosen, actual solution ideas can be developed. In doing so, the chosen frames serve as guidelines and criteria for assessment. This is a process of exploration, in which solutions are embodied and their rationale is tested. The purpose of this search is to arrive at sufficiently detailed and tested solution ideas. Often the goal is to generate a set of alternative acceptable solutions. The designers may be individuals, homogenous teams or heterogeneous groups, who use their professional and general knowledge to develop alternative solutions. It is at this phase that designers navigate among all the knowledge items in their design spaces, and possibly try to push their boundaries. Typically there are many connections between this tier and the previous framing tier.

5.4.4 Tier 4: Design Proposal

After a rigorous search, resulting in one or more design solutions that have been tested and shown to be viable, one solution is selected (or synthesized), which is then detailed up to a specified level. This solution is subject to cycles of evaluation and refinement, and to the approval of other stakeholders, such as clients and users.

Unless the design effort stops at this phase (for example, if there is no budget for its execution), the design is then materialized and put to use. We do not distinguish here between different categories of design products, such as completely new ones, redesigns, or variants of the same prototype.

5.4.5 Tier 5: Post-design—Lessons Learned

Section 5.2 emphasized the importance of the post-design phase, in which new knowledge is derived based on feedback from the use of the designed artifact. In addition to such feedback, which is practical for the most part, new theoretical knowledge may emerge if the design of the entity refers designers to new territories. A new building type may be defined (for example, a few decades ago a community center was a new notion), or a product within a known category is transformed with new properties that surpass what was hitherto known (for instance the first folding bicycles). In extreme cases, novel designs may bring about paradigm shifts and new norms, and they certainly may have an impact on the development and use of new technologies, as well as on regulations and laws.

5.4.6 Central Axis: Insight to Impact

We have looked at the five tiers of the design space; it may be appropriate to claim that they represent design expertise. Each of these tiers is an entity of its own, but they interact and rest on each other in many flexible ways. However, in addition to expertise creativity is also a hallmark of design almost by definition: every designed entity is new, at least in some respects. Can a designer (or team) come up with a design that does not only fulfill its purpose but is also considerably novel, based on expertise alone? Some say that this is possible, and a host of training programs and techniques aimed at enhancing creativity have been suggested (e.g., Higgins 1994; VanGundy 1988), many of which, like Triz and brainstorming, are not unique to design (for an insightful overview, see Smith 1998). Others are critical of 'creativity techniques' and claim that if at all used, they are but incarnations of design methods (e.g., Laakso and Liikkanen 2012). As such, creativity methods pertain to the expertise portion of the design space, in one or more of the bottom tiers.

In this paper we associate creativity not with any component of expertise, but with the axis of insight-impact. Insight is not divorced from knowledge, but it does not emanate from it directly. An insight may be a vision concerning the design solution, but it may also pertain to the users, or to the kind of search that should be undertaken. It requires no explicit rationale and is not predictable in advance; it is sudden and may surprise even the designer him or herself. Insights are closely related to intuition and imagination: "Working together intuition, (recognition), and imagination give rise to insight, the quintessential phenomenon of breakthrough creativity, the eureka moment, the sudden flash that brings new light to what previously lay in darkness" (Ogle 2007, p. 205). Burnette (2013, p. 10) explains that an insight has a neurocognitive basis whereby "selective attention and imaginative exploration influenced by the affective state of mind and relevant and useful knowledge loosely coupled to the primary focus of the thought" may result in insight. But, he adds, "resolving a problematic situation does not end with insight. Rather, the solution must be implemented and executed... and assimilated into knowledge" (Burnette 2013, p. 11).

In practice, an initial insight mobilizes various components of expertise and guides the activities in the design space. If successful, the insight is recognizable in the final design proposal and finds its way into the lessons learned at the phase of post-design (if not supported by knowledge in the design space, an insight may be lost). The impact the design has once implemented is a function of the quality and novelty of that design; the quality is often the outcome of expertise but the novelty is closely related to the initial insight. Therefore we discern the independent insight-impact axis, around which expertise allows the development of a design artifact. Every new design requires at least a modest insight at the outset. Otherwise, the design may be a solution to a problem but it may hardly qualify as a new design artifact. Obviously, there is a great difference between powerful insights that lead to important breakthroughs of a historical dimension, which Boden (1994) classified as belonging to the H-creativity category, and lesser insights that result in so-called P-creativity. For our purposes here the magnitude of creativity is of no consequence. What is of importance is the fact that the design space may be seen as consisting of the creativity insight-impact axis, and the tiers of expertise that are activated in the process of designing to reach the best, most functional, most novel and most influential outcome possible in the short and long terms alike.

5.5 Conclusions: Integrated Comprehensive Design

Design takes place within a design space. What is or should be the content of the design space is not a fixed given; it is for the designer or the design community to determine. At present the design space is limited to the task-dependent search for a solution, usually based on a collection of possible solutions, both known and unknown (to be discovered). There is a lot to be gained from expanding the scope of the design space to encompass the pre-design and post-design phases, which are currently separate entities, not well integrated into the design process. By being part of the design space it is estimated that prior knowledge and stances (pre-design) would play a more dominant role in design and would also be more easily transformed by it, thus being updated frequently. Likewise, the assessment of the actual performance of a design, once it is realized, could be better utilized to update knowledge, which could become part of the pre-design phase of a future design task. The inclusion of pre and post-design in the design space contributes to the level of expertise with which design assignments can be handled.

Novelty and creativity are in great esteem today and are recognized as driving the success of new products, beyond the functional quality of such artifacts. Economically speaking, novelty is an essential component of competitiveness in the market. Therefore many 'creativity techniques' are practiced in industry in an attempt to augment creative design thinking. This paper claims that the drive for creativity should be seen as an integral part of the design space and not something that is "appended" to it. It is recognized that creativity is not part of expertise, although it does of course rely on knowledge. Therefore, creativity is seen as an axis leading from insight to the potential impact of a design entity, and this axis is perpendicular to the facets of expertise and runs through them.

To model the expanded design space that embodies both creativity and all levels of expertise, a useful metaphor presents itself, that of the five-tier Japanese pagoda. This pagoda is the sturdiest of structures, as it handles with unprecedented agility the forces that run through the structure when submitted to even the severest of blows. The various components of expertise, as they are practiced along the design process, are likened to the five tiers of the pagoda. The insight-impact axis in the proposed design space is likened to the pagoda's central pillar. The comprehensive design space that is portrayed in this model should be as sturdy as the pagoda that serves as a metaphor for it, and should handle even the most difficult, messy and unstructured design tasks with confidence and efficacy.

Acknowledgement The author wishes to thank Toshiharu Taura and Yukari Nagai for organizing the Nara Design Research Leading Workshop, which in addition to being an excellent discussion forum, has also acquainted the participants with the intriguing and inspiring five-tier pagoda.

References

Boden MA (1994) What is creativity? In: Boden MA (ed) Dimensions of creativity. MIT, Cambridge, pp 75–118

Burnette C (2013) Intuition, imagination and insight in design thinking. http://www.academia.edu/3737350/Intuition_Imagination_and_Insight_in_Design_Thinking. Accessed 2 Dec 2013

Dorst K, Cross N (2001) Creativity in the design process: co-evolution of problem-solution. Des Stud 22(5):425–437

Goldschmidt G (1997) Capturing indeterminism: representation in the design problem space. Des Stud 18(4):441–455

Goldschmidt G (2012) A micro view of design reasoning: two-way shifts between embodiment and rationale. In: Carroll JM (ed) Creativity and rationale: enhancing human experience by design. Springer, London, pp 41–55

Goodman N (1978) Ways of worldmaking. Hackett, Indianapolis

Habraken NJ (1985) The appearance of the form. Awater Press, Cambridge

Hegedüs Á, Horváth Á, Ráth I, Varró D (2011) A model-driven framework for guided design space exploration. In: Proceedings of 26th IEEE/ACM international conference on automated software engineering, Lawrence, KS, November 6–12, 2011, pp 173–182

Higgins JM (1994) 101 Creative problem solving techniques. New Management Publishing Company, Winter Park

Laakso M, Liikkanen LA (2012) Dubious role of formal creativity techniques in professional design. In: Duffy A, Nagai Y, Taura T (eds) Proceedings of the 2nd international conference on design creativity, Glasgow, September 18–20, 2012, pp 55–64

Maher ML, Poon J, Boulanger S (1996) Formalizing design exploration as co-evolution: a combined gene approach. In: Gero JS, Sudweeks F (eds) Advances in formal design methods for CAD. Chapman and Hall, London, pp 1–28

Nakahara K, Hisatoku T, Nagase T, Takahashi Y (2000) Earthquake response of ancient five-story pagoda structure of Horyu-Ji temple in Japan. In: Proceedings of the 12 WCEE 2000: 12th world conference on earthquake engineering, Auckland, New Zealand, January 30–February 4, 2000, vol 2, Earthquake engineering in practice. http://www.iitk.ac.in/nicee/wcee/article/1229.pdf. Accessed 29 Nov 2013

Newell A, Simon HA (1972) Human problem solving. Prentice-Hall, Englewood Cliffs

Ogle R (2007) Smart world; breakthrough creativity and the new science of ideas. Harvard Business School Press, Boston

Rosenman MA, Gero JS (1993) Creativity in design using a design prototype approach. In: Gero JS, Maher ML (eds) Modeling creativity and knowledge-based creative design. Lawrence Erlbaum, Hillsdale, pp 111–138

Saxena T, Karsai G (2010) Towards a generic design space exploration framework. In: Proceedings of the 10th IEEE international conference on computer information technology, Bradford UK, June 29–July 1, 2010, pp 1940–1947

Schön DA (1983) The reflective practitioner. Basic Books, New York

Schön DA (1984) Problems, frames and perspectives on designing. Des Stud 5(3):132–136

Schön DA (1987) Educating the reflective practitioner. Jossey Bass, San Francisco

Schön DA (1991) Designing as reflective conversation with the materials of a design situation. Keynote talk, Edinburgh conference of artificial intelligence in design, June 25, 1991. http://www.cs.uml.edu/ecg/pub/uploads/DesignThinking/schon-reflective-conversation-talk-design-games.pdf. Accessed 5 Nov 2013

Sloman A (1995) Exploring design space and nitch space. In: Proceedings of the 5th Scandinavian conference on AI, Trondheim, May 1995. IOS Press, Amsterdam

Smith GF (1998) Idea-generation techniques: a formulary of active ingredients. J Creat Behav 32(2):107–133

VanGundy AB (1988) Techniques of structured problem solving, 2nd edn. van Nostrand Reinhold, New York

van Langen PHG, Brazier FMT (2006) Design space exploration revisited. AIEDAM 20(2):113–119

Wall K, Shea A (2013) Post-occupancy evaluation of a mixed use academic office building. In: Håkansson A, Höjer M, Howlett RJ, Jain LC (eds) Proceedings of the 4th international conference on sustainability in energy and buildings 2012 on sustainability in energy and buildings, SIST 22. Springer-Verlag, Berlin, pp 501–510

Westerlund B (2005) Design space conceptual tools – grasping the design process. In: Proceedings of In the making, Nordes – the Nordic design research conference, Copenhagen, May 29–31 2005

Woodbury RF, Burrow AL (2006) Whither design space? AIEDAM 20(2):63–82

Zimring CM, Reizenstein JE (1980) Post-occupancy evaluation: an overview. Environ Behav 12 (4):429–450

Chapter 6
Motivation in Design as a Driving Force for Defining Motives of Design

Hernan Casakin and Shulamith Kreitler

Abstract The chapter introduces a new approach to social and personal motivation for creativity grounded in the cognitive orientation (CO) theory, and discusses its relation as a major drive to design motive in the pre-design and post-design phases. The instrument for assessment of motivation for creativity—the cognitive orientation questionnaire of creativity (COQ-CR)—is described, and applied in a study about social and personal motivations of creativity of design students in architecture and engineering. The COQ-CR allows characterizing the structural and thematic composition of motivation for creativity across different design disciplines. Implications for design practice and design education include operationalizing the COQ-CR for understanding the underlying connections between motivation for creativity and design motives of designers belonging to different disciplines.

Keywords Architecture • Behaviour • Cognitive orientation • Creative design • Creativity • Engineering • Motivation • Motive • Post-design • Pre-design

6.1 Introduction: The Problem

In order to understand better the essence of design creativity, it is necessary to explore the design phases that occur not only during the problem solving activity, but also those that precede and follow it. These involve focusing on the pre-design and post-design phases. While the pre-design phase is characterized mainly in terms of explicit requirements and goals for the new products (e.g., Simon 1981), the post-design phase can be distinguished by the experiences, feelings, and perceptions that consumers or users express with regard to the design products (e.g., Taura 2014). The relations between these phases, however, are not always clear. Certain designs are produced by considering given specifications or goals, but without taking into

H. Casakin (✉)
Ariel University, Ariel, Israel
e-mail: casakin@ariel.ac.il

S. Kreitler
Tel Aviv University, Tel Aviv, Israel
e-mail: krit@netvision.net.il

© Springer Japan 2015

T. Taura (ed.), *Principia Designae – Pre-Design, Design, and Post-Design*,
DOI 10.1007/978-4-431-54403-6_6

account feelings expressed by users with regard to such designs. In other cases, designs are developed on the basis of the inspiration of the creative designer, while disregarding the demands of the user. This problem becomes even more critical in domains combining functional with artistic aspects, such as architecture, where social considerations are not always regarded as important as the personal ones. Thus, more research is needed to understand the relationship between the pre-design and the post-design phases, wherein the notions of motive and motivation of the designer are expected to play an important role.

Nagai and Taura (2009) argued that motivation and motive are two powerful components responsible for achieving the involvement of creative processes in design. Whereas a motive can be defined as an aim for acting in a certain way, motivation can be viewed as the engine necessary in order to activate, trigger, as well as to maintain and bring to completion a design motive. In other words, design motives are a necessary constituent of design motivation. For this reason, gaining insight into the structure and dynamics characterizing the motivation of the designer can contribute to understanding the underlying driving forces affecting his or her motives of design, in particular those that are conducive to design creativity. In this study we consider creativity as a cognitive act that integrates both the social and personal motives. Consequently, we aim at dwelling on the notion of design motivation, and its relation to design motive in the pre and post design phases.

We based the study on the Cognitive Orientation (CO) theory (Kreitler and Kreitler 1982; Kreitler 2004), which defines motivation as a function of a set of belief types, themes, and groupings identified as relevant for the development of creativity. The CO Theory is one of the more developed and comprehensive theories of motivation, including both a broad conceptual structure, and an assessment method with a well-established empirical basis. The major theoretical proposition is that behavior in all domains—including design and creativity—is a function of motivation and performance, whereby motivation is conceptualized in terms of beliefs. These are cognitive structures that represent particular themes, which are relevant for the behavior in question and may be organized into main groupings. In this study, we analyze design motivation by focusing on external (social motivation), and internal (personal motivation) aspects, which complement each other.

Understanding how these two kinds of motivation are assessed, is critical for learning more about design motives. Thus, we aim at gaining a deeper insight into the main aspects characterizing internal and external motivation for design creativity by analyzing two groups of students in the architecture and engineering design domains. We explore whether each group may differ in the relative salience of social and personal motivations, and we discuss how motivation may be connected to design motives, and what may be the implications thereof for design education. It is expected that the study will contribute to clarifying the relation between design motivation, design motives, and their connections to the pre-design and post-design phases.

6.2 Motivation in Design Creativity

Design activity consists of the production of outstanding, innovative, valuable, and unexpected outcomes for the sake of promoting progress and development in society. Design is therefore concerned with the particular attitude orienting towards creative thinking, which requires the ability to perceive reality from unorthodox viewpoints and to challenge established norms (Chakrabarti 2006). In addition to talent and special skills that have long been known to enable creativity, in the last decades it has become increasingly clear that motivation is a most influential factor affecting creativity. Some studies provided evidence about the significance of motivation in the creative process (Collins and Amabile 1999; Runco 2005). Motivation can be intrinsic or extrinsic (Amabile 1985; Kaufman 2002). Whereas the former is characterized by pleasure or satisfaction derived from performing the task itself, the latter is concerned with getting a reward that is external to the task. Motivation for creativity, and in particular intrinsic motivation, was shown to play a fundamental role in design (Chakrabarti 2010; Krippendorf 2004; Kröper et al. 2010). However, Nagai and Taura (2010) noted the difficulty in assessing intrinsic motivation in real time design processes.

Recently, Kreitler and Casakin (2009) investigated the function of motivational disposition for creativity in architectural design. Their study, which was grounded in the Cognitive Orientation theory (See section below), showed that motivation for creativity has a robust predictive power with regard to a large number of variables assessing different aspects of design creativity. In a comparative study that focused on design creativity and education, Casakin and Kreitler (2008) found significant differences between more creative and less creative students with respect to some motivational themes. The more creative students scored high in themes such as willingness to use talent to achieve originality, delve into the unknown, freedom to apply individual criteria, and readiness to make efforts and invest in the design task.

To date very few studies were carried out to investigate the influence of motivation on creative design achievements, and therefore more research is needed. Regardless of the important influence of motivation in creative design domains like engineering and architecture, designers are generally unaware of its effect, and only infrequently consider the motivational disposition aspect in the evaluation of creativity. This is often the case in most design studio activities, in which design creativity is promoted or assessed without considering the personal motivations related to it. Remarkably, a similar attitude toward motivation can be identified in regard to the role of motivation for cognitive acts in general, such as perception, memory, or thinking (Kreitler and Kreitler 1987a, b). An explanation may be that cognitive acts appear to be so natural that they seem to be activated without any need for motivation. Another reason may be that the outcomes of motivation in cognitive acts are not instantly manifested.

6.2.1 Motivation as a Trigger for Motive in Creative Design

A distinction can be made between motivation defined as "a desire to do; interest or drive" (Collins English Dictionary 2013), and motive, referring to "the reason for a certain course of action". A motive is a major object of interest that can be used as an argument in favour of, or a justification for doing something. Both, motivations and motives can be internal (personal), i.e., related to the inner reality of the self, or external (social), i.e., concerned with an external reality that can embrace social, or environmental aspects (e.g., Taura 2014).

Motives and motivations are interrelated, so it is not possible to carry out certain acts for whatever reason, without the existence of a relevant wish or goal. For example, the motive for designing a house is intimately related to the beliefs and understanding of the designer with regard to how people should dwell. Nevertheless, the relation between the motives for generating certain creative outputs, and the underlying motivational disposition is not symmetrical. Thus, the same motivational disposition may serve or implement different design motives. In contrast, different motivations can be triggered by similar motives.

While motives are expressed intentions to act or do something in a specific manner, motivations are generally unconscious (Freud 1957). So designers are generally unaware of the motivations affecting the implementation of their design motives. This may be due to the fact that it is easier to refer to the design motives than to the design motivations. This applies both to design practice and education. For example, design briefs are generally based on a programme with a list of functional needs, rather than on the explicit desires of the designer. Moreover, observations from the design studio show that teachers and students mainly focus on the creative outcomes and the motives leading to it, disregarding thereby the underlying motivations for engaging in creative acts. Lack of awareness of the motivational disposition of the designer may have negative consequences since, on the one hand, it discourages approaching motivation in a systematic manner, and on the other hand, it hinders the potential development of his/her creative skills. Ignoring motivational disposition might also limit the capacity to understand why certain decisions are taken instead of others.

6.3 Cognitive Orientation Theory and Method: Motivation and Motive

In this study, the conceptual and methodological approach grounded in the Cognitive Orientation (CO) theory (Kreitler and Kreitler 1982; Kreitler 2004) was used for exploring motivation for creativity in design, and its relation to design motive. The CO theory is a cognitive-motivational approach that deals with the identification, understanding, prediction, and change of behaviours in different domains, such as cognitive, emotional and motor (Kreitler and Kreitler 1976, 1987a, b).

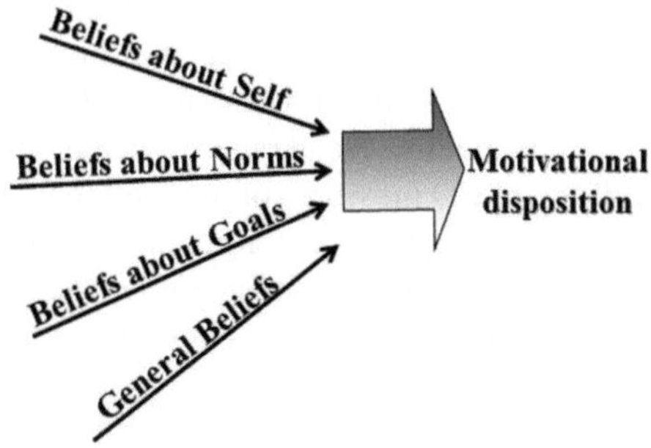

Fig. 6.1 Motivational disposition as a combined product of beliefs

A fundamental principle tenet of the CO theory is that an output is a function of a motivational disposition and a behavioural program that implements it. The motivational disposition is defined as the combined product of beliefs of a specific form and particular contents to which the beliefs refer. In terms of form, the beliefs refer to the following four basic types: (1) beliefs about oneself, which express information about oneself, such as one's lifestyles, activities, or feelings (e.g., 'I am curious', 'I use to play tennis'); (2) beliefs about rules and norms, which denote ethical, social, aesthetic, and other rules and standards (e.g., 'People should learn to respect others', 'Ladies first'); (3) beliefs about reality and others (general beliefs), which are concerned with information about others and the environment (e.g., 'Buildings should be sustainable', 'Users know what they need'); and (4) beliefs about goals, which refer to actions or states wished or not by the individual (e.g., 'I want to be successful', 'My desire is to travel around the world').

Regarding the contents, the beliefs are concerned with themes dealing with the meanings of the studied behaviour, which was identified and defined by a standard procedure of in-depth stepwise interviewing validated by empirical testing (Kreitler and Kreitler 1982). Accordingly, the beliefs represent themes that do not refer directly to the behaviour, but reflect the meanings underlying it. This ensures their significance for the behaviour under consideration (Kreitler 2004).

Further, directionality and strength are the main characteristics of the motivational disposition. Directionality controls the activity towards which the motivational disposition is oriented, whereas strength is defined by the number of belief types and the range of themes within each belief type orienting toward the particular activity (See Fig. 6.1).

The motivational disposition plays an overwhelmingly important role in regard to creativity. However, for a creative act to occur it is necessary to consider two further factors: the behavioural program that implements the motivational disposition in practice, and the trigger which was called above "motive". A motivational disposition cannot be activated in a void. Motive is the objective of the act, or the task that the individual engaged in a potentially creative act undertakes to perform, such as "designing a park", "planning an urban area". Tasks or projects of this kind trigger the activity of designing, which—depending on the designer's motivational disposition—may turn out to be more or less creative.

The cognitive model in which the CO theory is grounded proposes that the process of development of the motivational disposition is unconscious and unintentional, and not necessarily rational. Previous studies assessing CO motivations embracing a broad range of behaviours and domains of application showed the validity of CO measures of motivational tendencies (Kreitler and Kreitler 1987a, b, 1988a, 1994). Further predictions provided by the application of the CO theory were also confirmed for cognitive acts such as planning and problem solving (Arnon and Kreitler 1984; Kreitler and Kreitler 1987a, b, 1988b).

6.3.1 The CO Questionnaire: Assessing Motivation in Design Creativity

The motivational disposition can be assessed by means of a CO questionnaire. A questionnaire of this kind is constructed on the basis of a matrix which includes four columns representing the four belief types and rows representing the themes or the groupings of the themes, so that each cell presents a statement that is a belief referring to a particular theme. The CO questionnaire consists of statements, each with four response alternatives: very true, true, not true, and not at all true. The participant is requested to check one of these responses in regard to each statement. In the CO questionnaire all the items referring to each belief type are presented consecutively, in random order. The Cognitive Orientation Questionnaire of Creativity (COQ-CR) assessing the motivation for creativity includes 384 items which are classified into four parts, each dealing with one of the four types of belief (i.e., 98 items for beliefs about self, 91 for general beliefs, 98 for beliefs about norms and 97 for beliefs about goals). The items referred to 79 themes that were identified in prior studies as representing attitudes related to creativity in different domains (e.g., Not satisfied with any achievement but seeking more; Creating something personal; Devoting time and effort to society; Curious to learn a lot about every domain). A shorter version of the COQ-CR has been recently finalized.

The COQ-CR was developed and applied in a number of studies that centred on creativity, such as devising innovative uses for energy (Margaliot 2005), solving engineering problems (Casakin and Kreitler 2009), and architectural problems (Kreitler and Casakin 2009). Applying the COQ-R questionnaire in design has several advantages. The major ones are that (i) it was constructed on the basis of a theoretically-driven procedure that supports the assumption that the themes are appropriate for creativity; (ii) it covers a broad range of contents that represent a profile of creativity; (iii) it has been tested empirically in domains demanding creative thinking; (iv) it is not subject to rational and conscious considerations on the part of the respondents; and (v) it has a wide applicability in regard to creative tasks in various domains.

Assessing the relations of motivational predisposition to design creativity represents a challenge, and has important implications for practice and education. Such an approach may contribute to theorizing about the role of motivation in design creativity, and its relation to the design motive.

6.3.2 CO Theory: Social and Personal Motivation for Creativity

Table 6.1 presents the 11 groupings of the original 79 themes of the COQ-CR. The groupings were obtained by means of cluster analysis of the theme, and were supported by factor analysis (Casakin and Kreitler 2009; Kreitler 2013). The 11 groupings representing the themes constituting the motivational disposition can be organized in terms of two main categories corresponding to: (i) external (social) motivation aspects, and (ii) internal (personal) motivation aspects (See Table 6.1). Each of these can form a CO cluster which is manifested as a vector. It may be hypothesized that under certain circumstances these two CO clusters of social and personal motivation may be combined into one single vector and function together in a balanced form or unbalanced form with one or the other playing the dominant role, or they may contradict each other and result in a conflict situation.

In order to gain insight into the application of the CO theory and the internal and external motivational predispositions in design creativity, a comparison between architectural and engineering groups was carried out (see Sect. 6.4).

6.4 The COQ-CR: Comparing the Domains of Architecture and Engineering—Social and Personal Motivation

In a previous study, Casakin and Kreitler (2009) explored the motivational disposition for creativity of 52 students of architectural design, and 60 students of engineering design. The comparison was carried out to evaluate the stability of the conceptual structure of the CO questionnaire across samples, and to determine similarities and differences between the architectural and engineering design groups. Thereafter, *t*-tests were run to compare the means of the COQ-CR for the four belief types, and the 11 groupings in the two groups. The 11 groupings were also factor analyzed.

It is remarkable that similar factors were found in the two groups of students, further validating the structure of the CO questionnaire for creativity. In both groups the first factor referred to aspects related to personal motivation: it refers mainly to themes concerning the self—its uniqueness, development and expression. The second factor dealt mainly with themes concerned with social motivation: openness to the environment and the society, but without risking inner directness.

Table 6.1 The 11 groupings of the themes of the COQ-CR organized according to internal and external aspects

Social motivation aspects	
External groupings	Examples of themes
Contributing to the society and the community	Make something important and significant, even if it does not contribute to self-promotion; Feeling that one can promote the general welfare; Devoting time and effort to society; Readiness to invest a lot to help people
Being receptive to the environment, absorbing from the environment	Extracts something from the environment even if it offers only a few stimuli (openness to the environment); Absorbing from the environment as much as possible, not selectively; Curious to learn a lot about every domain
Freedom in functioning	Does not need a rigid framework of rules defining the situation and the conditions; Unable to function according to the instructions of others; Functions by intuition; Need of freedom in thinking and acting; Acting because he/she wishes and not because he/she ought to
Functioning under conditions of uncertainty	Liking ambiguity and uncertainty; Liking to take risks; Liking jobs in which not everything is clear; Functioning even if he/she cannot control every detail of the process
Non-functionality	Able to work even if sees no immediate benefit; Does not believe that every idea can be implemented in practice; Compromising in regard to practicality and functionality; Readiness to act even if functionality is not clearly stated, or not clearly requested from the start
Personal motivation aspects	
Internal groupings	Examples of themes
Self-expression	Expressing emotions outwardly; Creating something personal; Expressing thoughts, views, and skills; Externalizing feelings; Being loyal to own feelings and ideas; Speaking with others about oneself
Self development	Investing in oneself and developing oneself; Taking advantage of opportunities for promotion and learning; Developing skills; Not satisfied with any achievement but seeking more
Emphasizing one's uniqueness	Experiencing one's uniqueness; Feeling that he/she has unique talents; Developing and highlighting one's uniqueness; Understanding things in his/her own way; Being different from others; Seeing things differently from others; Making original things
Emphasis on inner world	More interested in what takes place within oneself than in what occurs outside and in others; Making efforts to learn about him/herself; Feeling contradictions within the inner world; Emphasizing the importance of fantasy

(continued)

Table 6.1 (continued)

Personal motivation aspects	
Internal groupings	Examples of themes
Inner directedness	Making efforts to succeed in circumstances in which others tend to fail; Lack of support from others does not affect self-confidence or self-esteem; Clarity about one's goals
Demanding from oneself	Does not withdraw in the face of difficulties; Striving for perfection, getting to the level of excellence one determines for oneself; High demands from oneself; Investing without limits; Able to renounce comfort and pleasure

The results suggest that personal motivation themes were identified as being more important for design creativity than those dealing with social motivation. Most importantly, despite the architectural and engineering groups may differ in their nature, a common profile of motivation for creativity based on both personal and social motivation was shared by both groups.

Additional results showed differences in two of the four belief types that are supportive of design creativity. Scores for the beliefs about the self, and beliefs about goals were found to be higher in the architectural group than in the engineering one. It seems that the motivation for creativity in the architectural group is mainly driven by beliefs concerned with personal motivation. The resulting outcome can be described as a combination of the actual self-image, and the aspiring or ideal self-image as represented in one's personal goals. It is interesting that the groups did not show any difference with regard to their social motivations, which included general beliefs about how people and things actually are in reality, and how things should be, as reflected by rules and norms. These results highlight the differential structure of motivation for creativity in the students of architecture and engineering.

Further differences between the architectural and engineering groups were observed in the motivational structure for the groupings supporting design creativity. The motivational disposition of the architectural students was characterized by the salience of inner-directedness, inner world, and development of the self, stressing one's talent, uniqueness and self-expression, all of which can be considered to be part of the personal motivations of the designer. In contrast, the motivational pattern of the group of engineering students was characterized by absorbing from the environment, being receptive to the environment, as well as to demanding from oneself, despite potential difficulties. Accordingly, the aspects of social motivations were more dominant in this group.

A more detailed analysis was carried out in order to understand the complex thematic structure of motivation for creativity that was prominent in each group of students. Findings showed that the architectural group was characterized by a need of autonomy, valuing the creation of something personal, a sense of uniqueness, being guided by intuition and emotions, and emphasizing the self over the external environment and the others. Accordingly, all of these themes are concerned with personal motivations of the designer. On the contrary, the profile for the engineering group was defined by openness to the environment, manifested in a specific interest to learn and absorb from it as much as possible, even under conditions in which the environment provides few stimuli. In addition, they had social and political concerns, as well as a tendency to demand from themselves, and strive for perfection even at the cost of distancing from the others. Most the themes reflected an environmental and social concern.

6.5 Discussing Motivation and Motive in Creative Design

Analysing the thematic structure of motivation for creativity at different levels of complexity by means of the COQ-CR, allow identifying the more general differences between groups of designers belonging to different disciplines, as well as differences between individuals that may belong to a same design field but may have dissimilar motivational tendencies.

Applying the COQ-CR for understanding the internal/personal, and external/social motivations of designers from different disciplines can contribute to gaining a deeper insight into the drive leading to their design motives, and the role that these play in the pre-design and post-design phases. As a result, it may be possible to strengthen links between these design phases, and reduce potential mismatches between the aims of the designer and the needs of the user. By knowing what might be the internal and external motivations of a designer according to his or her professional background and individual differences, it would be possible to learn whether these are balanced, or unbalanced due to the dominance of one or another aspect over the other. For example, when the internal motivations are stronger than the external ones, the link established between external motivations and social motives might be weak or disregarded. In this case, it would be possible to make designers more conscious of specific external motivations related for example to the needs of the users with regard to a specific design, or to the environment itself. Consequently, it may be necessary to make designers aware of how users interact with their designs, and what are their resulting experiences, wishes, perceptions and aspirations. Whereas social motives acquire relevance mainly in the post-design phase, gaining insight into them can have a powerful effect not only in the external motivations of the designer, but also the internal ones. Since motivations are active mostly in the pre-design phase, they can affect the formation of personal motives related to the definition of design needs and requirements for future designs. In other cases in which design situations are totally new and social needs are unknown, there could be a missing connection between the existing designs and the designer's motives for a new design (Nagai and Taura 2009). Therefore, it might be critical in the pre-design phase to help designers clarify what are the internal motivations influencing their personal aims for design action.

However, the connections developed between motivation for creativity and design motives are not free from controversy. Thus, a revolutionary design whose initial conception is based mainly on the internal motivations of the designer, in a post-design phase might be considered as creative but inadequate for the users' needs. In contrast, a conventional design based mainly on external motivations with a strong concern for actual requirements, in a post-design phase might be positively assessed by the users for its value, although it would not be considered as highly innovative. This paradox demonstrates that despite the substantial roles played by motivation and motives of the designer in the pre- and post- design phases, more research is needed to gain further insight into this topic.

6.6 Conclusions and Implications

This work is a first step in the study of motivation for creativity, and its relation to design motives in different design disciplines. The Cognitive Orientation theory was used to define motivation for design creativity. Similarities and differences regarding motivational tendencies between architectural and engineering students were analysed.

Regarding the contents, the CO theory suggests that motivation for creativity is a complex and multifaceted concept. A measure of such complexity enables numerous combinations of themes to define motivation for creativity in designers belonging to different disciplines, and with different profiles. Consequently, a group of designers may have the same level of creativity motivation, but may differ in the structure and the themes of their motivation. Since motivation for creativity is a necessary factor for activating and bringing to completion design motives, the manner in which he motivational disposition is structured may have a direct effect on how design motives are dealt with. The major implication of the study is that the COQ-CR is a reliable tool that enables exploring, examining and understanding the motivational disposition for creativity and henceforth considering it in any study or project of design creativity.

The stated implications are highly relevant also for intervention programs of education seeking to promote creativity in design domains such as architectural or engineering, while at the same time considering personal and social motivations. Using the CO theory, which is a theoretical framework for predicting and changing behavioural manifestations, can contribute to promote individual motivations for creativity in each specific design domain in a straightforward way. An advantage of the CO theory over other cognitive-motivational approaches is that it enables to identify the motivational profile of a designer, and introduce changes by means of a standard procedure of intervention originated by the CO theory. Intervention programs relying on this approach could be applied by strengthening, or increasing the number of internal or external beliefs and the related themes associated to a desired behavioural manifestation (Kreitler and Kreitler 1990). Using the CO approach can endow design educators with a suitable tool for promoting motivation for design creativity, and for clarifying its relation to design motives in the different phases of the design process. The pedagogical application of this instrument and the intervention procedure can be adapted to the needs of the different design disciplines.

References

Amabile TM (1985) Motivation and creativity: effects of motivational orientation on creative writers. J Pers Soc Psychol 48(2):393–399

Arnon R, Kreitler S (1984) Effects of meaning training on overcoming functional fixedness. Curr Psychol 3:11–24

Casakin H, Kreitler S (2008) Motivational aspects of creativity in students and architects: implications for education. In: Clarke A, Eyatt M, Hogart P, Lloveras J, Pons L (eds) Proceedings of the international conference on engineering and product design education, 4–5 September, Barcelona

Casakin H, Kreitler S (2009) Motivation for creativity in architectural design and engineering design students: implications for design education. Int J Technol Des Edu 20:477–493

Chakrabarti A (2006) Defining and supporting design creativity. In: International design conference – DESIGN 2006. Dubrovnik, Croatia, pp 479–486

Chakrabarti A (2010) Motivation as a major direction for design creativity research. In: Taura T, Nagai Y (eds) The first international conference on design creativity ICDC2010. Kobe, pp 49–56

Collins English Dictionary (2013) Complete & unabridged, 10th edn. http://www.collinsdictionary.com/. Accessed 10 Apr 2013

Collins MA, Amabile TM (1999) Motivation and creativity. In: Sternberg RJ (ed) Handbook of creativity. Cambridge University Press, New York, pp 297–312

Freud S (1957) The unconscious. In: Strachey J (ed and Trans) The standard edition of the complete psychological works of Sigmund Freud, vol 14. Hogarth, London, pp 166–204 (original work published 1915)

Kaufman J (2002) Dissecting the golden goose: components of studying creative writers. Creat Res J 14(1):27–40

Kreitler S (2004) The cognitive guidance of behavior. In: Jost JT, Banaji MR, Prentice PA (eds) Perspectivism in social psychology: the Yin and Yang of scientific progress. American Psychological Association, Washington, DC, pp 113–126

Kreitler S (2013) The structure and dynamics of cognitive orientation: a motivational approach to cognition. In: Kreitler S (ed) Cognition and motivation: forging an interdisciplinary perspective. Cambridge University Press, New York, pp 32–61

Kreitler S, Casakin H (2009) Self-perceived creativity: the perspective of design. Eur J Psychol Assess 25(3):194–203

Kreitler H, Kreitler S (1976) Cognitive orientation and behavior. Springer, New York

Kreitler H, Kreitler S (1982) The theory of cognitive orientation: widening the scope of behaviour prediction. Prog Exp Pers Res 11:101–169

Kreitler S, Kreitler H (1987a) Plans and planning: their motivational and cognitive antecedents. In: Friedman SL, Scholnick EK, Cocking RR (eds) Blueprints for thinking: the role of planning in cognitive development. Cambridge University Press, New York, pp 110–178

Kreitler S, Kreitler H (1987b) The motivational and cognitive determinants of individual planning. Genet Soc Gen Psychol Monogr 113:81–107

Kreitler S, Kreitler H (1988a) The cognitive approach to motivation in retarded individuals. In: Bray NW (ed) International review of research in mental retardation, vol 15. Academic, San Diego, CA, pp 61–123

Kreitler S, Kreitler H (1988b) Horizontal decalage: a problem and its resolution. Cogn Dev 4:89–119

Kreitler H, Kreitler S (1990) Cognitive primacy, cognitive behavior guidance and their implications for cognitive therapy. J Cogn Psychother 4:155–173

Kreitler S, Kreitler H (1994) Motivational and cognitive determinants of exploration. In: Keller H, Schneider K, Henderson B (eds) Curiosity and exploration. Springer, New York, pp 259–284

Krippendorf K (2004) Intrinsic motivation and human-centred design. Theor Issue Ergon Sci 5(1):43–72

Kröper M, Fay D, Lindberg T, Meinel C (2010). Interrelations between motivation, creativity and emotions in design thinking processes – an empirical study based on regulatory focus theory. In: First international conference on design creativity ICDC2010. Kobe, pp 97–104

Margaliot A (2005) A model for teaching the cognitive skill of melioration to pre-service science teachers in a college for teachers. Unpublished doctoral dissertation, Bar-Ilan University, Ramat Gan, Israel

Nagai Y, Taura T (2009) Design motifs: abstraction driven creativity. Jpn Soc Sci Des 16(62):13–20

Nagai Y, Taura T (2010) Discussion on direction of design creativity research (Part 2). Research issues and methodologies: from the viewpoint of deep feelings and desirable figure. In: Taura T, Nagai Y (eds) First international conference on design creativity ICDC2010. Kobe, pp 9–14

Runco MA (2005) Motivation, competence, and creativity. In: Elliot AJ, Dweck CS (eds) Handbook of competence and motivation. Guilford, New York, pp 609–623

Simon H (1981) The sciences of the artificial. MIT, Cambridge

Taura T (2014) Motive of design: roles of pre- and post-design in highly advanced products. In: Chakrabarti A, Blessing LTM (eds) An anthology of theories and models of design: philosophy, approaches and empirical explorations. Springer, London

Chapter 7
Affording Design, Affording Redesign

Barbara Tversky

Abstract Thinking quickly overwhelms the mind, and the mind expands into the world. The mind creates a range of cognitive tools to represent thought. Sketches are among the most prevalent and productive ways to make thought visible to self and others and to promote creative thought. Sketches, like diagrams, map the elements and relations of ideas to elements and relations on the page. They entail abstraction and allow ambiguity. Their ambiguity creates possibilities, it allows exploration, inference, and discovery of ideas, some of the reasons they are so useful in design as well as other domains. Objects and the surroundings that provide contexts for objects can also serve as tools for thought. Like sketches, objects and surroundings provide visual feedback, but unlike sketches, objects and environments provide tangible feedback, feedback from interactions with the body as well as the eyes of the user. Whereas design begins with ideas and goals that need embodiment, redesign begins with a specific problem embodied in a person in a place at a time. Solutions often come from reuses of old objects. Problems and reuses of designed products inspire implicit conversations between designers and redesigners.

Keywords Ambiguity • Constructive perception • Creativity • Design • Diagram • Map • Redesign • Sketch

7.1 Thinking with Things

Creativity seems impossible to study. First, we don't know what it means. And second, whatever it means, it goes on in the mind, invisibly. Whatever it is, creativity is not just wild ideas, unusual associations, weird combinations. Any dumb machine can do that, probably more wildly than people. Although it may sound paradoxical, creativity is constrained. Creative solutions may indeed appear

---

B. Tversky (✉)
Columbia Teachers College, 525 W. 120th St., New York, NY 10027, USA

Stanford University, Stanford, CA, USA
e-mail: btversky@stanford.edu

© Springer Japan 2015

T. Taura (ed.), *Principia Designae – Pre-Design, Design, and Post-Design*,
DOI 10.1007/978-4-431-54403-6_7

wild, unusual, and weird, but they suit needs, fit goals, satisfy constraints. Thus, creativity has two components: divergence and convergence.

Divergent thinking expands, each connection leading outwards to many more. Convergent thinking contracts, drawing connections across the divergent ones. Either of these processes quickly exhaust the mental workspace. There is an ancient solution: expand the mental workspace to include a physical workspace, put it in front of the eyes instead of behind them. One of the most flexible cognitive tools is a sketch or diagram. Just as a stick lengthens the arm, a sketch broadens and expands the mind.

Sketches externalize thought, enlarge the mind, force abstraction, provide a playground for exploration of new ideas, make ideas visible to self and others. Sketches are a natural tool for designers as they map the parts and relations of what is to be designed onto the space of a page. Sketches can be tentative and vague, expressing and suggesting many possibilities. Think of the amount the mind can hold at one time as a *mindful*. A sketch can hold several mindfuls, allowing designers to see far more than they can imagine, allowing designers to integrate mindfuls. As they emerge, sketches give feedback to designers: is something missing? do the parts fit together? does it make sense? how could it be otherwise? how could it be better? Sketches can be enhanced, especially in conversation, with talk and gesture, primarily iconic and deictic gestures. And, a boon for researchers, sketches provide data for those who wish to study design.

Externalizing ideas onto paper has benefits beyond extending the mind. Paper provides a space, a space we can be organize and reorganize to suit our needs, just as we organize physical spaces. Just as the physical organization of the spaces we inhabit inside and out serve our memory and our actions, the organization of a sketch serves memory and anticipated action.

Externalizing thought is by no means limited to design of objects and spaces. Sketching fosters design of the abstract as well as design of the concrete, sketching to imagine business organizations or processes as well as sketching to imagine objects and their uses. The space of a page, like the space of a room or a field, invites structuring. In the abstract case, a set of ideas, a page invites structuring and representing those ideas as parts and relations. Space itself carries meaning. Some of those meanings are expressed in gesture and words. Common gestures, like thumbs up, common expressions like "top of the class" or "falling into depression" are indicative of a rich set of associations to the vertical. Up is *more*, *higher*, *stronger*, most likely because going up requires overcoming gravity, which in turn requires age, strength, health, and power. Nothing as dramatic as gravity distinguishes the horizontal dimension, which is uniform in all directions around us. However, biology and culture confer some degree of asymmetry to the horizontal dimension. Most people in the world are right-handed, and there is some support for the influence of handedness at least on judgments of value, associated with greater dexterity and fluency of action. But cultural inventions, notably writing, seem to confer broader and stronger correspondences. In languages written left-to-right, *left* is *earlier*, *more powerful*, *first*, and motion that goes rightwards is *faster*. The opposite holds in languages written right-to-left (Tversky 2011a). The marks on a page as well as location on a page also carry meaning. Larger marks indicate larger thing. Larger marks also attract more attention, and size is used to indicate importance. Closer in paper space suggests closer in conceptual space. Sketching or diagramming ideas make abstract ideas concrete, hence easier to conceptualize. Putting ideas on a page promotes inferences based on spatial reasoning, proximity, direction, distance as surrogates for more difficult abstract reasoning. Crucially, for designers of any kind, sketches invite not just organization but also reorganization, easy trial and error (Tversky 2011a).

Design is nothing if not full of paradoxes and contradictions. Divergent and convergent thinking are at its core. The new and the old. Letting the mind wander and focusing it. Similarly, any tool, cognitive or physical, has disadvantages as well as advantages and many of the disadvantages are their very advantages. Sketches make the abstract concrete, and thereby force concreteness, promoting concrete spatial thinking. They structure and organize, thereby forcing premature specificity Putting ideas on a page can enable thought but also can constrain thought in ways that may have nothing to do with the problem at hand. Interacting effectively with sketches and diagrams requires expertise, to make the concrete abstract again, to see abstract implications in concrete spatial relations, to reconfigure configurations.

7.2 Designing Sketches: Sketches Reveal Thought

Sketches are exactly that, sketchy, schematic. They give an idea of what a designer has in mind, but they are inexact and incomplete, blobs for entities, like buildings, rough lines for relationships, like paths. They omit detail that hasn't been worked through or that is not relevant. In emphasizing the essential aspects of the design they not only simplify, they may exaggerate and distort. In so doing, they can reveal the underlying representation or model the designer has in mind. The schematic nature of diagrams also suggests possibilities, invites inferences, and enables revisions.

7.2.1 Sketch Maps

Maps are one of the commonest uses of sketches, by everyone. Although maps, even sketch maps, could be analog, people's sketch maps of regions or routes typically straighten curved roads and make streets more parallel and perpendicular. Just as they don't pay much attention to angle, they don't pay much attention to distance, often compressing large distances with few choice points or other information and enlarging small ones containing many choice points or other information. They reduce large complex environments not just in size but also in detail, including only detail of relevance (e.g., (Fontaine et al. 2005; Lynch 1960; Taylor and Tversky 1992; Tversky 2001)). Route maps get further simplified; they typically include only the relevant path (Tversky 2011b; Tversky and Suwa 2009).

People's mental representations of routes consist of links and nodes, where the links are typically paths or streets, and the nodes are typically intersections or landmarks, usually accompanied by names. Significantly, this same structure, and the same elements, underlies people's descriptions of routes and environments (Taylor and Tversky 1992; Tversky 2011b; Tversky and Suwa 2009). This suggests a phenomenon that appears across many domains: the same mental model drives descriptions and depictions.

How do we know all this? That is, how do we study the sketches? We collect a large number from individuals under standard conditions and code them. The coding is both a priori—we have some ideas of what we expect to find and look for them—and a posteriori—in looking through the sketches people produce, we see phenomena that seem interesting and code those as well. The coding has to be reliable; that is, two scorers following the same coding scheme have to agree. The coding takes into account what isn't represented as well as what is represented. For what is represented, it takes into account how it is visualized. Roads, for example, are usually represented as single or double lines; the landmarks, intersections or buildings or distinctive features of the environments, that appear are represented as vaguely-shaped blobs or simply as names. In the distortion of angle and distance, we compare the real angles of streets and distances to what participants produce. Similarly for omissions, we compare what people put in their maps to what they could have put in. We count all those cases and compute statistics on the counts. Then we generalize, making small leaps from the data: to the notion of links and nodes; to the distortions of angle and distance; to the essentials, from what is and isn't.

7.2.2 Systems Design

A current project is examining how master's students in a course in design of information systems create and use sketches in the service of design (Nickerson et al. 2008). The design of information systems is elusive because systems consist of concrete objects like computers and servers that have a real configuration in real space. However, what is critical is not the real configuration but a functional conceptual one; how the objects are networked. Euclidean distance among components is not important; their network connectivity is. What is particularly hard for students to grasp is a *bus*, a module or group of objects that are mutually interconnected, but may have unique links into and out of the group. A bus is essentially a hierarchical structure, but represented on flat paper. Throughout the course, students solved a series of design tasks. Most design tasks gave a description of a system to be designed and asked students to make a sketch of the system. They were also asked to make inferences from the system, for example, the set of shortest paths through the system. Reasoning about paths of information flow is crucial to system design. Both students' diagrams and their inferences showed that especially early on, they had problems with the bus structure, indicated by faulty sketches as well as errors of commission, specifically, listing paths that were not shortest paths. We called this a *sequential bias*, to treat all connections as linear and ordered. Systems diagrams bear superficial resemblance to more familiar route maps, which are appropriately interpreted sequentially. The inference errors of omission indicated a second bias, a *reading order bias*. Students tended to list shortest paths in reading order, from upper left to lower right, and to miss the later paths. Importantly, students who did master the bus in their sketches made more correct inferences from them. In both cases, students are bringing previous habits for interpreting marks on pages to design; the previous habits impede interpretations of systems design sketches. The data supporting these conclusions depended on characterizing the diagrams as well as the inferences and comparing the two.

Other work on diagrams corroborates these finding. Research on describing environments represented as sketch maps revealed a reading order bias (Taylor and Tversky 1992). When the environments had a natural beginning, like an entrance, people tended to start their descriptions at the entrance. But when environments did not have a natural beginning, people began their descriptions at the upper left. Another striking example is a resistance to diagram cycles as circles, instead preferring to represent them linearly. When students are asked to make a representation on paper to show the seasons or to show the water cycle, most students make linear arrays rather than circular ones (Kessell and Tversky 2007). That is, they see the processes as having a beginning, middle, and end rather than endless repetition. Put together, the findings suggest that students bring habits from everyday spatial reasoning and interactions with text and with simpler diagrams, such as maps, to the creation and understanding of sketches of more complex and subtle ideas. The implication is that creation and interpretation of sketches should be carefully taught.

7.2.3 Ideas

It has become clear that the core structure of a broad range of sketches is nodes and links, where nodes are typically entities, concrete or abstract, places or ideas, and links are the connections among them. Another, more abstract, way of putting it is that the nodes are nouns and the links are predicates; nouns and predicates of course are the basic elements of language. In fact, the node-link structure is also commonly used for abstract ideas, where the nodes are virtual entities, concepts, and the links are relations among them. The applications are too many to list, from concept maps to encourage order in ideas in school children to flow charts to decision trees to visualizations of the web and Facebook networks. The ubiquity of node-link diagrams suggests that they reflect core qualities of human thought. Two nodes and a link is a minimal diagram; a subject and a predicate a minimal sentence.

7.2.4 More Than Nodes and Links

Nodes and links do not exhaust what sketches do and can do. As noted, sketches add language, where depictions won't do. Sketches add icons for objects, buildings, and the like. Sketches add a range of schematic graphical forms, lines are one, but that group also includes arrows, boxes, blobs, brackets, boxes, and more. These especially have meanings that are readily interpreted in context from their geometric or gestalt properties (Tversky et al. 2000). Lines connect, indicating relations; arrows are asymmetric lines, indicating asymmetric relations. Boxes contain and separate. Blobs are vague shapes, indicating 2- or 3-D objects that make take more definite shape. Because these schematic forms readily take on meanings that vary with context, they are useful and used across a broad range of sketches and diagrams.

7.2.5 Sketches Consolidate Learning

As noted, sketching provides a natural way to represent a set of organized thoughts and relations among thoughts. This is a fortiori the case for thoughts and relations that have physical counterparts, like instructions to operate or assemble something or explanations of how something is structured or behaves. Although teaching of these myriad sets of ideas is often accompanied by diagrams and sketches, learning is typically assessed verbally. The correspondence of meaning to words and sentences is purely symbolic. There are many benefits to using sketches and diagrams in teaching; there should be comparable benefits to students in learning. We set out to see if that is true (Bobek and Tversky 2014). In two studies in actual classrooms, junior high students were taught a STEM system, in one case, a bicycle pump from interaction with a pump, and in the other case, chemical bonding from a multi-modal video. After learning, students in the chemical bonding lesson were tested, though students in the bicycle pump condition were not. Then students were divided into two groups. Half were asked to provide traditional verbal explanations of the phenomena and half were asked to provide visual explanations. The explanations were coded for the information expressed, especially information about structure and behavior, process, causality. In general, more of that information appeared in the visual explanations, some of that in the words and symbols that accompanied the visual explanations. After completing the explanations, students were again tested. Note that no additional learning took place after the original lesson. For the bicycle pump, students of low spatial ability who produced visual explanations outperformed those who produced verbal ones. This is a simple device, so it is not surprising that there were no differences in the students of high spatial ability. In the more complex case of chemical bonding, all students performed better after creating explanations then immediately after learning, so the simple act of explaining without new study or information augments learning. And students of all ability levels performed better after creating visual explanations than after creating verbal ones. Their sketched diagrams differed widely, some even had metaphors, like sharks, hands, boiling pots. The power of creating visual explanations—sketches—seems to come from the natural mapping of parts and relations among parts of the systems to the page. That mapping provides a check for completeness, a check for coherence, and a platform for making inferences from structure to behavior, process, causality.

7.2.6 *Sketches Foster Collaboration*

Sketches, as we have seen, make ideas public, the invisible visible. The externalization of thought onto paper, where it can be inspected and reinspected, eases communication. Sketches are more permanent than speech, and ideas structured on paper are more interpretable than ideas conveyed in words. They provide a physical model that can be internalized to a mental one.

Sketches also allow easy embellishment with speech and gesture, to further enhance communication. Because much essential information is there in the sketch, that information doesn't have to be conveyed in awkward, clumsy speech. It can be pointed to or gestured on, accompanied by deictic terms like "this," or "that way," or "from here." In one project, pairs of participants were given a hypothetical map of a situation on campus after an earthquake, where some roads were closed, and injured gathered at special places. Participants were asked to draw a new map showing a route that would rescue the largest number of injured citizens as efficiently as possible. Some pairs worked over the same map, so they could gesture and see each others' gestures. Other pairs were separated by a curtain, so they could easily hear each other, but not see each other. The pairs who could work together over the same map produced far better new maps in less time, and were much happier with the interaction (Heiser and Tversky 2004; Heiser et al. 2004). The main kinds of gestures were pointing at landmarks (0-D), tracing paths (1-D), and sweeping areas (2-D). These gestures served dual roles, a deictic role of pointing to or indicating particular places and an iconic role, showing the shape of something. The dependent measures here are the quality of the sketch maps, as rated by blind judges, the subjective reports of the participants, and the analyses and comparisons of the gestures and the accompanying speech for both groups of participants, co-present and remote.

In this case, the gestures on the sketches were in the service of the design of a route. There is every reason to believe that gestures and speech on sketches will facilitate communication and joint design of other things. In fact, using an externalization of thought in the form of a sketch would seem to be especially effective for more abstract ideas, which may be harder to grasp without a stable external representation that provides some structure to the ideas.

7.2.7 *Sketches for Design: Promoting Inference and Discovery*

So far, we have examined how sketches represent thought and how they convey thought. Each is a step in creative thinking. But creativity also depends on going beyond what is in a sketch, in making inferences, seeing implications, altering and recombining elements, and more. Going beyond the information in a sketch is key, and expertise helps. There are numerous examples. The addition of arrows to diagrams of mechanical systems, such as a car brake or bicycle pump, turns interpretations from structural to functional. Without an arrow, people describe the spatial relations among the parts. With arrows, people describe the causal action of the system from start to finish (Heiser and Tversky 2006). Data presented as bars induces inferences of discrete comparisons whereas data presented as lines induces inferences of trends (Zacks and Tversky 1999). The students in the class in design of information systems who drew correct sketches also made correct inferences from their sketches (Nickerson et al. 2008). Getting the sketch right gets the thinking right.

This especially holds in design, where designers use successive sketches to hold design conversations (Goldschmidt 1991, 1994; Schon 1983). Designers put skeletal ideas on paper, and then reflect on what they've drawn. Sometimes they see new things in their diagrams, things they might not have intended. Visual patterns repeated, or overall configuration; these are perceptual inferences. Designers also see, make inferences about, functional aspects of what they are designing. They may see the flow of traffic or the light changing throughout the day or seasons. They may see bottlenecks or unused open spaces. These processes can be studied using protocol analyses. As they design, designers report their thoughts out loud; their thoughts are categorized and coded, and the temporal progression and interlinking of thoughts are analyzed (Goldschmidt 1991, 1994).

In our own work, we asked expert and novice architects to design a museum under certain constraints (Suwa and Tversky 1997). They worked (with pleasure, for the most part) for about an hour, usually producing a series of sketches, gradually refining their designs. Afterwards, they were shown a videotape of their sessions and asked to report why they made every pencil stroke. This retrospective protocol analysis allows the designers to focus completely on the design task, without interrupting themselves, a more natural situation. However, it runs the risk of forgetting and reinterpretation on the part of the designers. Nevertheless, the protocols have indeed provided valuable data.

In an early analysis, the information in the protocols was categorized by topic: emergent properties such as shape and size; spatial relations such as "connected" or "near" or "configuration;" functional properties such as views or circulation; and conceptual knowledge. Typically, there were longs strings of comments on one of these topics, and then a shift of focus. The architecture students were not consistent in what attributes drove the shifts of focus or the continuing strings. In contrast, for the architects more than the students, focus shifts were characterized by both spatial and functional relations. Similarly, for the architects, the connections between the continuing strings tended to be functional much more than for students. The dual implications are that students are not adept at inferring the functional information from the sketches, and that architects are adept at integrating the perceptual and the functional when focusing to a new aspect of the design, and to continue thinking about the new aspect in high-level functional ways.

Further analyses of the data provided a window on the origin and consequences of unintended discoveries, that is, things designers see in their own sketches that they had not intended. This phenomenon points to one of the key advantages of diagrams (Suwa et al. 2001). Designers drew them with one thing in mind, but later, when inspecting their sketches, see new relations and properties and implications once the diagram is in front of their eyes. They may see patterns in the locations of structures that present a potential theme that will unite and integrate the components, a perceptual discovery. They may see patterns of traffic or play of light, conceptual discoveries.

Sketches are deliberately ambiguous early in the design process. Designers do not want to be locked into particular structures and shapes and spatial relations at first. The ambiguity fosters unintended discoveries as well as flexibility of rearrangement. In fact, the protocol of one experienced designer showed clearly that after he perceived a new arrangement of parts, he was more likely to make an unintended discovery, and after an unintended discovery, he was more likely to regroup the parts.

This positive cycle, of unintended discoveries generating reinterpretations and reinterpretations generated new unintended discoveries seems general and productive. In a partner study, students were asked to generate as many interpretations as possible for a set of ambiguous, suggestive sketches. Those who spontaneously adopted a strategy of mentally reorganizing the parts of the sketch generated significantly more new interpretations than those who didn't. Other strategies were not as effective. This positive cycle also suggests simple ways to foster design creativity, or possibly creativity in other domains: to encourage and practice rearranging, reconfiguring parts of a whole. More abstractly, this would translate to changing the relationships among the components, and then interpreting them.

Generating new interpretations is an essential part of creativity, the divergent part. But especially in design, creativity needs also to be focused to an end or an outcome. That requires convergent thinking. Further research investigated that, again using the paradigm in which students are asked to generate new interpretations for ambiguous sketches (Suwa and Tversky 2003). In that study, students, both designers and non-designers, were first given a remote associates test, requiring then to find a relationship between distant word concepts. This test requires focusing on the aspects of the concepts that are abstract or aspects that connect the concepts. Students who scored high on the remote associates test also produced more interpretations for the ambiguous figures. These two predictors of generating new ideas, reinterpretation and remote associations reflect the two sides of creative thinking: reorganization and connection, divergence and convergence. In the service of design, the interaction of this dual set of processes was termed *Constructive*

Perception, the active reorganization of perception in the service of a design aim. Like other design processes, especially sketch-dependent ones, it is an iterative one: looking, seeing new arrangements, perceiving new relations, eliciting new design roles. Constructive perception is a pair of abilities, but a pair of abilities that can be trained.

These related projects have integrated three methods in the study of design creativity: retrospective protocol analysis that reveals the hidden design deliberations and decisions; generation of new interpretations of ambiguous figures; individual differences; expertise. Two factors foster the creative use of sketches, expertise and ability, and these trade off, one can compensate for the other.

7.2.8 *What Can Be Learned from Sketches?*

Sketches are a cognitive tool, closely related to diagrams, visualizations, and graphics. Here, a variety of projects were reviewed, demonstrating a variety of roles for sketches in design: revealing thought, fostering collaboration, and promoting inference and discovery. Although the present focus is design, sketches, and even more, their tidied-up sister, diagrams, are effective in learning, teaching, thinking, and communication in many domains of life, getting from one place to another, putting something together, operating equipment, understanding scientific phenomena, sharing and revising ideas (Tversky 2001). What makes sketching in design special is ambiguity. In navigation, in construction, in instruction, in communication, clarity is essential, and diagrams are often even distorted in order to make essential information clear. In design, especially in creative design, ambiguity is essential, as ambiguity encourages many interpretations and reinterpretations.

Sketches are a natural tool used in design of the concrete and the abstract. They expand the mind, make thought visible, and allow inspection and reinspection, promoting inferences and fostering recombinations. These processes and more have been revealed from studying how sketches are produced and used by novices and experts.

7.3 Design and Redesign

Whereas design begins with a goal, redesign begins with a problem. Whereas design begins with an idea, redesign begins with a thing, or things. Whereas design is abstract, redesign is concrete. Whereas design is decontextualized, redesign takes place in situ. Whereas design is done by professionals, redesign is done by ordinary people. Whereas design is for the masses, redesign is for an individual. Whereas design is for many situations and many times, redesign is for the here and now. Redesign is opportunistic. When Kohler's ape could't reach the bananas, he tried jumping. No luck. In a flash of insight, he pile up a bunch of boxes strewn around, climbed up, and got the banana. Countless coat hangers have been bent to reach something under a bed; countless paperclips have been bent to hold a torn garment together, countless objects into doorstops. Adrian Tomine's comics cover of the *New Yorker* February 25, 2008 is a poignant story of redesign. The 9 panels depict a writer writing a book, an editor okaying it, manufacturing the book, purchasing the book, reading the book, putting it in the trash, where it is retrieved by a homeless person, the redesigner, who burns it to stay warm.

But this view of design and redesign is short-sighted. Just as designers call their interactions with their sketches "conversations," design and redesign can be viewed as conversations between designers and users. Design is iterative, products are constantly being redesigned to accommodate users and their inventions. In the era of rapid apps, redesigners become designers; the flash for the camera on a smartphone becomes a flashlight, the gravity sensor becomes a level, the speaker a white-noise generator to put babies to sleep. When those unanticipated uses flourish, they affect the next design. Design starts with a disembodied idea, a goal, redesign starts with a concrete situation embodied in a person at a time in a space. The human imagination is remarkable but not omniscient. At its core, the human mind is associationistic, and the associations that arise from disembodied ideas and goals are necessarily different from the associations that arise from embodied problems encountered in multifaceted times and places. That conversation between designers and redesigners is necessary: designers cannot imagine all the situations and uses of their creations and redesigners cannot imagine all the creations that could solve their problems.

Acknowledgements Gratitude to the following grants for partial support of some of the research: ONR NOOO14-PP-1-O649, N00014011071, and N000140210534, NSF REC-0440103, NSF IIS-0725223, NSF IIS-0855995, and NSF IIS-0905417 and the Stanford Visual Analytics Center.

References

Bobek E, Tversky B (2014) Creating visual explanations improves learning. In: Bello P, Guarini M, McShane M, Scassellati B (eds) Proceedings of the 36th annual conference of the cognitive science society. Cognitive Science Society, Austin, TX

Fontaine S, Edwards G, Tversky B, Denis M (2005) Expert and non-expert knowledge of loosely structure environments. In: Mark D, Cohn T (eds) Spatial information theory: cognitive and computational foundations. Springer, Berlin

Goldschmidt G (1991) The dialectics of sketching. Creat Res J 4:123–143

Goldschmidt G (1994) On visual design thinking: the vis kids of architecture. Des Stud 15:158–174

Heiser J, Tversky B (2004) Characterizing diagrams produced by individuals and dyads. In: Barkowsky T (ed) Spatial cognition: reasoning, action, interaction. Springer-Verlag, New York, pp 214–223

Heiser J, Tversky B (2006) Arrows in comprehending and producing mechanical diagrams. Cognit Sci 30:581–592

Heiser J, Phan D, Agrawala M, Tversky B, Hanrahan P (2004) Identification and validation of cognitive design principles for automated generation of assembly instructions. In: Proceedings of advanced visual interfaces'04, ACM, pp 311–319

Heiser J, Tversky B, Silverman M (2004) Sketches for and from collaboration. In: Gero JS, Tversky B, Knight T (eds) Visual and spatial reasoning in design III. Key Centre for Design Research, pp 69–78

Kessell A, Tversky B (2007) To forge straight ahead or go round in circles. Poster presented at the meetings of the Psychonomic Society, Long Beach, November

Lynch K (1960) The image of the city. MIT, Cambridge

Nickerson JV, Corter J, Tversky B, Zahner D, Rho YJ (2008) Diagrams as a tool in the design of information systems. In: Gero JS, Goel A (eds) Design computing and cognition '08. Springer, Dordrecht

Schon DA (1983) The reflective practitioner. Harper Collins, San Francisco

Suwa M, Tversky B (1997) What architects and students perceive in their sketches: a protocol analysis. Des Stud 18:385–403

Suwa M, Tversky B (2003) Constructive perception: a skill for coordinating perception and conception. In: Proceedings of the cognitive science society meetings

Suwa M, Tversky B, Gero J, Purcell T (2001) Seeing into sketches: regrouping parts encourages new interpretations. In: Gero JS, Tversky B, Purcell T (eds) Visual and spatial reasoning in design. Key Centre of Design Computing and Cognition, Sydney, pp 207–219

Taylor HA, Tversky B (1992) Descriptions and depictions of environments. Mem Cogn 20:483–496

Tversky B (2001) Spatial schemas in depictions. In: Gattis M (ed) Spatial schemas and abstract thought. MIT, Cambridge

Tversky B (2011a) Spatial thought, social thought. In: Schubert T, Maass A (eds) Spatial schemas in social thought. Mouton de Gruÿter, Berlin

Tversky B (2011b) Visualizations of thought. Top Cogn Sci 3:499–535

Tversky B, Lee PU (1998) How space structures language. In: Freksa C, Habel C, Wender KF (eds) Spatial cognition: an interdisciplinary approach to representation and processing of spatial knowledge. Springer-Verlag, Berlin, pp 157–175

Tversky B, Lee PU (1999) Pictorial and verbal tools for conveying routes. In: Freksa C, Mark DM (eds) Spatial information theory: cognitive and computational foundations of geographic information science. Springer, Berlin, pp 51–64

Tversky B, Suwa M (2009) Thinking with sketches. In: Markman A, Wood L (eds) Tools for innovation. Oxford University Press, Oxford, pp 75–84

Tversky B, Zacks J, Lee PU, Heiser J (2000) Lines, blobs, crosses, and arrows: diagrammatic communication with schematic figures. In: Anderson M, Cheng P, Haarslev V (eds) Theory and application of diagrams. Springer, Berlin, pp 221–230

Tversky B, Agrawala M, Heiser J, Lee PU, Hanrahan P, Phan D, Stolte C, Daniele M-P (2007) Cognitive design principles for generating visualizations. In: Allen G (ed) Applied spatial cognition: from research to cognitive technology. Erlbaum, Mahwah, NJ, pp 53–73

Zacks J, Tversky B (1999) Bars and lines: a study of graphic communication. Mem Cogn 27:1073–1079

Part III
Technology and Society of Pre-design, Design, and Post-design

Chapter 8
Towards Data Democracy Beyond Fukushima

Shuichi Iwata

Abstract Data-driven design procedures on materials, artifacts (engineering products) and environmental projects are comparatively studied in accordance with view of three loops for learning, namely, learning how to learn, changing the rules and following the established rules. Materials design procedures including the three loops by data-intensive ways are under development by taking advantage of a qualified database on materials so as to establish an exemplar for data-driven design in general with an explicit articulation. Design procedures of nuclear reactors have changed from data-driven design in the beginning to experience-based design of tacit knowledge, and standard-based design reflecting number of committed experts for design and development. This explanation of the nuclear reactor design is extended to show difficult issues about design procedures for environments which have not only one directional time dependent evolutional features as natural phenomena but also are full of human dimensions of about seven billion people with consequent uncertainties. As concluding remarks, an articulation to converge into productive cycles for the latter two design problems is given, extending procedures for data-driven materials design by introducing human dimensions for nuclear reactors and adding irreversible path dependent features and human dimensions for environmental issues.

Keywords Artifacts • Data-driven design • Discovery approach • Engineering products • Environmental project • Experience based design • Fukushima Daiichi • Materials • Nuclear reactor • Safety • Standard-based design

8.1 Exemplars in Materials Design

Design procedures, in general, are iterative converging procedures to obtain better solutions, getting new data, insights and visions, although they can be defined as an action to find mappings between a value space of users and a fact space on each

S. Iwata (✉)
Graduate School of Project Design, Tokyo, Japan
e-mail: s.iwata@mpd.ac.jp

© Springer Japan 2015

T. Taura (ed.), *Principia Designae – Pre-Design, Design, and Post-Design*,
DOI 10.1007/978-4-431-54403-6_8

object obtained by observations and/or measurements by analytic prolongation and singularity treatments to cover unknowns by knowns.

As the value space of users reflects many factors associated with users, it implies human dimensions anyway. Thus it has been discussed as general and popular subjects in humanities, philosophy, economics, engineering and so on. However the value space on materials can be almost equivalently described in terms of properties by correlating them to a value or function in a system where materials are used. And the fact space on materials can or should be described in terms of data, logics and/or physics following constituent principle: the basic constituent of matter are various kinds of identical particles, with causality, covariance, invariance and equi-probability principle (Ni 2014). Moreover in case of materials, the fact space can be associated with the structural space of structural data of the constituents with finite patterns and their derivatives endorsed by geometry, for example, crystallography, we can rewrite materials design procedures to find mappings between the fact space and the structural space which can be associated with the value space by so-called structure–property correlation.

On the basis of these frames, materials design under development can be summarized as procedures by data-intensive ways including three loops for learning, namely, learning how to learn (discovery approach to find better mappings through verification and evaluation on each model derived systematically from data (Villars et al. 2008)), changing the rules (LPF approach by linking data associated with structural data including first principles models and cluster variation methods) (Chen et al. 1996) and following the established rules (total quality control of data by such established rules as geometry, symmetry, valence and so on), and we call the strategic combination of three loops as data-driven design. As far as the patterns to link associated data are finite, we can reach design solutions anyway by adding data step by step confining singularities at boundaries of solution spaces as shown in Fig. 8.1. Combinations and arrangement of constituents are classified by finite patterns (yellow dots and blue dots in Fig. 8.1 for example) which are associated with derived numerical values computed systematically from attribute values of constituents. Finding the best computation corresponds to finding the best mapping between the value space yellow dot dominant or blue dot dominant regions here and the fact space described by the attribute values of constituents. This method of discovery is explained as inverse problem solution for models, which is equivalent to finding design methods. In short it is a process of learning how to learn something from data systematically based on a creation of quasi-universal set prepared as one window of high quality almost exhaustive data.

Further design procedures of materials to meet engineering requirements are carried out step by step by taking into account property changes due to microstructural variations, defects, aging, service conditions and so on adaptively. And they are processed in design procedures of engineering products.

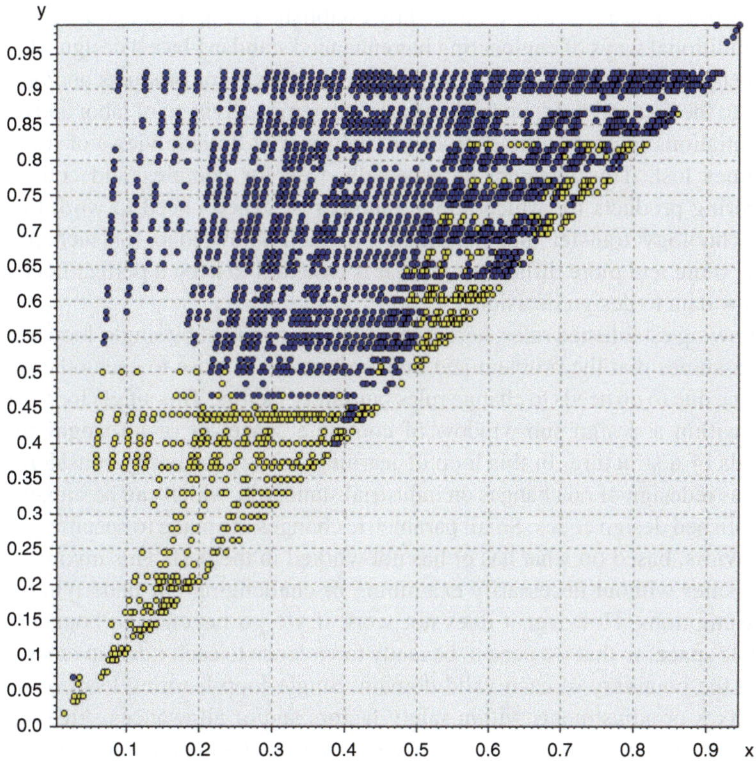

Fig. 8.1 Separation of 2,318 binary systems into compound formers and non-formers (Villars et al. 2008)

8.2 Design of Complex Artifacts

Nuclear reactors can be categorized as most complicated and sophisticated engineering products, and over one million parts are integrated to convert about 200 MeV fission energy to eV level thermal energy. Many prototypes of nuclear reactors at first were designed with a limited number of best experts when data-driven design was applied with respect to nuclear properties of constituent nuclei and intrinsic properties of materials. Traditional redesign approaches did not work in the beginning as there were no prototype engineering products for nuclear applications. However if we need to design nuclear reactors by taking into account such structure-sensitive properties as corrosion, strength, ductility, irradiations effects, aging, other degradations, we need to shift to experience-based design. In the beginning we did not have any design experience to builton. Full experimental, theoretical and computational surveys of all possibilities were the way to find potential candidates as a collection of prototypes, which have been summarized into a collection of de facto standards for design, and finally endorsed industrial standards. It is effective and practical as far as we can control facts within a pre-determined design window defined by established design standards. Shifting to these traditional ways of engineering has enhanced standard-based design. However design engineers lose traceability before the complexity of products and/or in situ feeling to the fact and field. Specialization of expertise, division of labor and domain differentiations increase efficiency of production, but holistic views of design are sometimes lost. This tendency is especially true for complex and complicated engineering products for which a large number of experts need to work together. And technology transfer and inheritance of tacit knowledge for such products become more and more difficult, so that it is essential to keep a perfect traceability from raw data to design decisions.

Following established rules, here standards, is carried out by single-loop learning, which assumes that the problem and their solutions are close to each other in time and space due to no needs to change rules but safety factors. This single loop learning works within a design sub-window of complete space, or of no singularities as materials of a structure. In this loop of learning, we are primarily considering our design as redesign of no changes on industrial standards, which can be implemented as established design codes. Small parametric changes are made to specific practices or behaviors, based on what has or has not worked in the past. This involves doing things better without necessarily examining or challenging our underlying beliefs and assumptions. However it does not work if we go beyond the boundary of a complete space, so that we need to be ready to violence to each rules, in other words, beyond the boundary of each valid domain. Single-loop learning leads to making minor fixes or adjustments within safety factors and/or allowances. Are we doing things right in accordance with established industrial standards and regulations? Here's what to do—procedures or rules or industrial standards. Key points here are openness on where we are within a set of rules and actions what to do if else.

The 14-metre-high tsunami after the magnitude nine earthquake of 11 March 2011 severely damaged the Fukushima Daiichi Nuclear Power Plant of the Tokyo Electric Power Utility Company (TEPCO), in particular, the plant's principal power supply. The wave also took out the backup electricity sources, leaving four reactor buildings—including four pools containing spent fuel elements without power. Consequently the earthquake and tsunami caused the station blackout (SBO) by losing all power sources, which has resulted in a kind of "multiple organ system failure" of the Plant bereft of coolant. We had no light, we could not cool the reactor core and spent fuel pool actively, and we could not measure the exact status of the plant by taking advantage of a set of advanced instruments and tools to do so. Available tools to measure radiation doses have been given priority for the ongoing emergency and to workers health management. Without enough sensors and "powerful hands", careful and coordinated efforts have been continued with perseverance to make each unit more stable and safer in a step-by-step manner by TEPCO and other experts. We had no "CPU" and no "Intensive Care Unit" to support and restore damaged parts in the beginning. All these accident sequences was triggered due to differences of safety design guidance on the position of emergency power supply facilities.

In short, the nuclear reactor systems were damaged mechanically and chemically, but they shut down perfectly with respect to nuclear chain reactions. Therefore, the Fukushima Plant is different from the accident at Chernobyl. At Fukushima, ranging fuel burn-up (averaged for each fuel assembly) from 3,200 to 41,100 MWd/t, 1,498 fuel assemblies in Unit 1–3 reactor cores, 2,724 fuel assemblies in spent fuel storage pools are producing decay heat, and at high temperature a part of cladding materials zircaloy of the fuel assemblies had reacted with steam/water to produce hydrogen. Three buildings were severely damaged due to the hydrogen explosion, and explained less exploded where a blow-out panel worked properly. Radioactive materials have been released from containment vessels on several occasions after the Tsunami, and this was the result of deliberate venting in an attempt to reduce gas pressure in each containment vessels and/or spent fuel exposures due to dry out. It is said the radioactive level was too high for workers to open the valve for venting if they follow the conservative regulation. The radiation dose rate at one location between reactor units 3 and 4 was measured very high at 400 mSv/h, at 10:22 JST, on March 13th of 2011 and the dose rate level near the road of the two units was about 1 mSv/h at 13:30 JST on January 27, 2014 by the author's measurement.

Risks to natural disasters, levels of safety technology, social risks, political risks and consequent crisis are different from place to place, from company to company and from country to country. Therefore in introducing an engineering product, designers need to rewrite the set of standards to meet engineering requirements of their own organization or country by retroactively checking semantics of the standards up to a set of raw data. By such rewriting designers get a realistic image of each engineering product at the site where and when the product is in service. When introducing new safety guidelines and standards into one engineering product, for example, stress test to evaluate safety against terrorist attack, natural disasters and cyber attack we need to follow double-loop learning which leads to insights about why a solution works. In this loop of learning, we are considering our actions in the framework of our up-to-date scientific knowledge and our operating assumption as engineering actions. This is the level of process analysis where people become observers of themselves on the basis of their own holistic viewpoints, asking, "What is going on here and there? What are the patterns as a whole?" The so-called linear model from idea inspiration and intuition, prototyping, manufacturing, marketing, operation and maintenance and waste managements used to be modified in accordance with in situ requirements at each phase. We need insights to understand each unexpected pattern. We change the way-rules, standards, regulations and operational guidelines etc., and we make decisions and deepen understanding of our assumptions. Double-loop learning works with major fixes or changes, like redesigning an organizational function for new safety standards or structure of artifacts. The more complicated and complex a targeting artifact becomes, the more times we ask at each phase of design. We are checking continuously "Are we doing the right things? Here's why this works—insights and patterns are reported to generate big files adding additional complexities due to human-made combined uncertainties.

Mismatches of safety guidelines as was the case of earthquake/tsunami in US and Japan can be managed ideally by certain individuals with talents of grasping critical issues and making proactive actions against such problems. However it was not the case before complicated and complex engineering products as nuclear reactors, and the Fukushima Daiichi nuclear reactor accident happened. We need to develop a way to overcome this kind of issues about experience-based design anyway.

The next triple-loop learning involves principles, namely, what is nuclear safety, why we need to use nuclear energy, how to deal with radioactive wastes, how to mitigate climate changes and so on. The learning goes beyond scientific insight and patterns and to holistic contexts of economical, social, cultural, ethical and other aspects. The result creates a shift in understanding our context or point of view. We produce new commitments by many stakeholders and ways of learning to see how to decide what is right. This form of learning challenges us to understand how problems and solutions are related differently by committing stakeholders, even when separated widely by individuals in terms of value, time and space. It also challenges us to understand how our previous actions created the conditions that led to our current problems. The relationship between organizational structure and behavior is fundamentally changed because the organization learns how to learn. The results of this learning includes enhancing ways to comprehend and change our purpose, developing better understanding of how to respond to our environment, deepening our comprehension of why we chose to do things we do, and refining our procedure of why we select a procedure of decisions under uncertainties of each design issue and differences of committing stakeholders. It is an endless procedure but we need to repeat "How do we decide what is right? Here's why we want to be doing this"—principles overcoming differences there. After the Fukushima Accident, engineering guidelines which had worked at the single loop did not work at the double loop due to the mismatches of safety design standards revealed by the 2011 Tōhoku earthquake and tsunami. Radioactive materials released by the Accident have contaminated the land where many people had been living, and make the issue into the triple loop "How do we decide nuclear reactors are necessary in Fukushima? Why do we still want nuclear energy in Japan?" Triple loop learning is about learning how to learn. It has been carried out by different stakeholders without explicit definitions of design windows at the triple loop, namely, whether it is a discussion for the world, Japan, Tokyo, Fukushima or peoples who have been forced to be evaluated. The relation of the three loops are shown in Fig. 8.2.

From a viewpoint of engineering product, the single loop learning is applied to each scientific domain, namely, materials engineering, chemical processing, reactor physics and so on. Integration of such different domains are dealt with at the double loop learning to reach a holistic design solution. Public discussions used to be carried out at the triple loop learning where the single and the double loop learning included. Ironically we are easily thrown into a panic on accidents, disasters and crisis even if trained well. Additionally we have a limitation to understand a complex and complicated product perfectly and holistically in general. Therefore it is important to design engineering products, safety guidelines, regulations as

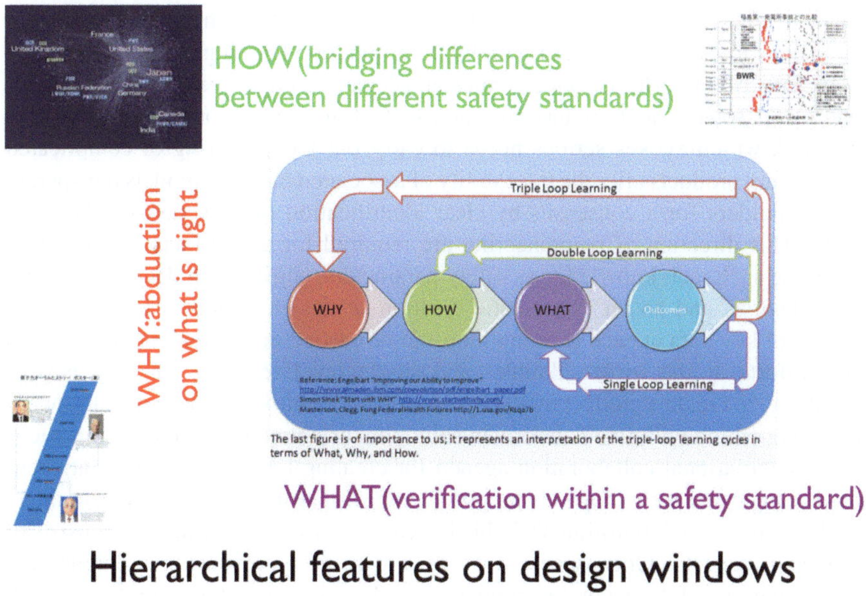

Fig. 8.2 Multiple loops to converge into design solution. Single loop learning on what, double loop learning on how and triple loop learning on WHY. Prepared by adding issues on the Fukushima nuclear reactor accidents after a scheme of learning (Debategraph 2014)

simple as possible for all committed experts to understand essentially important points. And it is also necessary to organize ad hoc design teams for these three loops and also ad hoc three operation teams again for these three loops. In this context we were prepared neither on safety standards at the double loop nor on emergency evacuations at the triple loop.

As for unknowns in science and technology there used to be debates in terms of values, risks and crisis, where risk evaluation, risk communication, risk managements and crisis managements have been endless issues. Depending on contexts, we have risk takers and risk rejecters, and silent customers of values. Guidelines and regulations on safety of artifacts, in general, have been or had been revised continuously on the basis of unexpected failures and accidents by the double loop learning as well as politics by stakeholders by the triple loop learning. In the evolving process of guidelines and regulations, experience based approaches have been articulated into model based approaches and data-based approaches endorsed by traceability, and others, namely, trial and errors as the double loop learning. Standards have been established as snapshots of such evolutions processes in engineering knowledge infrastructures in the double loop learning, but sometimes are under over-expectation regarded as an endpoint of the triple loop learning.

The nuclear reactors require a long time to become stable, and we are going forward step by step with trial and errors with incremental improvements. We are learning to derive solutions to take care the nuclear plant after the worst "Multiple organ dysfunction syndrome", but the worth fact is the consequent social big confusions to be carefully dealt with by the triple loop learning.

In case of Fukushima Daiichi nuclear accidents, cognitive gaps in evolving features of science and engineering, and locked-in explanations on technologies mattered, which evokes serious needs to carry out reengineering of complicated engineering products after the traceability of associated data and models transparent to stakeholders for a consensus by clear identification of each task on the three loops. This direction of reengineering are required for all artifacts of unknowns including such tacit knowledge as safety parameters, engineering decisions, know-how, theories and so on. Systematic creation of design windows by exhaustive listings of key data at each learning loop can be carried out incrementally. Beyond such efforts at the triple-, double- and single level learning we may have a possibility to create a product model by systematic iterations of the above discovery approach, which will give data-driven solutions for unknowns as data democracy.

An international collaboration agenda for creating a new paradigm "data centered science and technology" is proposed on health issues as triggered from many troubles due to the Fukushima Daiichi Nuclear Power Plant. The agenda consists of two main aspects: One aspect concerns a way to reorganize our engineering knowledge on safety beyond experience-based engineering, targeting to establish robust enough engineering against the next big risk and disaster, and the second aspect concerns how to create a human-oriented society of perfect individual care. Sharing of reliable scientific data is the key starting point for this big challenge. Challenges to realize data-centric individual cares and data- and scenario-driven collaborative design approaches together are expected to follow as remedies to get our future perspectives.

8.3 New Agenda for Environmental Project

Design of environments in principle requires a common mission shared by all, namely, about seven billion human beings, artifacts and natures, keeping a clear vision on climate and environmental changes evoked by human activities. And a common mission is to be derived or designed at the triple loop learning as is the case of Fukushima Accident although the committed people and stakeholders have been struggling to find solutions. By any means, we must be prepared against various risks, disasters and crisis based on not only reliable data and knowledge but also on restructuring our domain-differentiated organizations formed by domain-differentiated disciplines and knowledge into a tough, robust, flexible and basic organizations of perseverance.

In order to make the difficult triple loop learning fruitful, in parallel we need to establish robust engineering in design against unknown future risks, disasters and crisis based on our research from the biggest disaster during these 1,000 years. It is the way to make productive challenges on environmental projects of more difficulties in the future and requiring more vivid and wider time-space imaginations on longer term climate changes, on diversities of humanities even with some apparently cyclic patters as a planet in the solar system.

Lessons for the artifactual design from the Accident are

- Design mistakes by setting the emergency help diesel generator and power switches in the basement causing SBO at the single loop
- Missing inheritance of design knowledge about the blow-out panel from designers to maintenance technicians at the single loop and the double loop
- Missing of discussions on how to modify established regulations in accordance with the realistic evaluation on the accident cite at the triple loop

By using template of the above three loops for example, we nuclear experts need to explain our learned lessons explicitly to articulate the engineering, social, economical, environmental and human health issues that the Accident has evoked to present and future generations. It is crucial to recognize all the impacts are associated each other in complex manners and to see each fact in a holistic way, describing the fact by associated and relevant data and their explanations.

It is important for all to recognize that we were suddenly put into the triple loop learning and there are different observers, instruments, sampling methods and time of observations and time when observations are made. The reported raw data will have scatters and spikes and sometimes errors due to the ongoing complex event, so that it is important to evaluate the reported data independently and scientifically on the way of the single loop if possible with other scientists, and if possible by digital logic filters prepared by experts before. In our information society, various voluntary experts of different disciplines in the world can carry out evaluations of reported data, so that exaggerating biased reports and mistakes will be phased out gradually but its takes time without a set of standard reference data. An important prerequisite to make it happen is that we scientists need to communicate with each other by a common language beyond each scientific domain without using specialized technical terms Shibboleth, hopefully, through facts/objects and accessible scientific data in the beginning.

In case of calculated data, we need to know at least functionality and correctness of the program, valid range and domain, input data, computational parameters, accuracy and uncertainty of calculated data, and objectives of estimation, for example, best estimate or conservative estimate, phenomenological or first principles, deterministic or probabilistic, etc. So we need to be very careful in using calculated data without reference data and/or reference programs. Just a collection of calculated data without explicit explanations has brought confusions to the society also in this occasion. One lesson we have learned is we need to prepare reporting views for calculated data of importance.

It is not easy to control "Information Tsunami" after something happened as it passed rapidly through different mass medias and the Internet together with the strong images of hydrogen explosions as the consequences of severe core damages. It requires data commons to support the society at the level of the triple loop learning, and it needs to be prepared and maintained in a feasible way. Risk evaluation data are to be prepared in the way our society can use to protect before.

Our society has not trusted nuclear experts due to the bad unexpected sequences just after the Earthquake/Tsunami/Nuclear Accident, which has been rubbed into the mind of society repeatedly. It is a serious lesson for scientists. As the first step to get the future perspective on longer-term climate changes, it is important to carry out joint fact finding periodically by all committed experts and convert their findings into a set of lessons on design in general as public goods. Transparencies on the procedure to converge into a unique voice by scientists are the key in case of complicated issues. No one knows the right actions against climate changes with local fluctuations, disclosures of all data and knowledge are required to organize collaborative challenges by all people.

In short we do not know about the earth perfectly, so that we cannot predict the next earthquake/tsunami accurately enough with respect to the magnitude, time and place for preparation. What we can do now is to make our engineering for safety robust enough against any big risks in the future, and especially to recover the intrinsic resilience of people to protect themselves based on the available data by their own capacity to read data and capability to collaborate with colleagues.

Defense-in-depth and/or fault-tolerant design approaches do not work if we follow experience-based design staying within one specialized academic domain. We need to prepare an open environment-data commons ironically in preparation due to the Accident using radioactive materials as tracers, and the data commons will give us a precise view on the dynamics of environments, especially, water, soil and sea streams. Establishing robust enough engineering against the next big risk and disaster is a very challenging subject, but can be done by resolving all ideas into a set of knowledge and meta-knowledge. In this context "Fukushima" can be articulated as a strategic procedure to derive local and global control of a complex, heterogeneous, non-linear and dynamic multi-component environmental system by identifying the status of the system with limited data.

People who cannot protect themselves have suffered from natural disasters. Each individual has a fundamental human right to know his/her own risk rather than ambiguous indices of his/her place. EU Maastricht Treaty says "The absence of certainties, given the current state of scientific and technological knowledge, must not delay the adoption of effective and proportionate preventive measures aimed at forestalling a risk of grave and irreversible damage to the environment at an economically acceptable cost." So it looks more practical to carry out health care of each individual who may have a risk suffered from the release of radioactive materials, rather than of an unspecified large number of people with a poor statistics. This way of thinking is true also health issues by any means of environmental issues as PM2.5 although it is important to control the source term.

Scientific research is required to maintain the public health focusing on individual care, which may explore another dimension for the public care. In case of Chernobyl, a holistic approach by combining tele-medicine and human genome database has been under discussion for practical solutions, and it could be a good reference for any health issues.

Reliable scientific data and individual care policy can be the basis for reasonable solutions. In this context, calculated data requires careful evaluation on their reliability and we need to use calculated data in a proper way with reliable references. How safe is safe enough? It is a question concerning not only science but also sociology, which requires powerful and elaborate analysis on interplay of science and society back and forth in both qualitative and quantitative way.

The government's role of informing the public of the risks is not easy due to the prematured status of our society at the triple loop learning. The person to provide this information must speak with knowledge and authority taking into account uncertainties of the risks. Obviously there's a lot of concern about food and water associated health issues, and it seems uncertain about the pathways and contamination. Methods and/or procedures to evaluate risks and uncertainties are to be explained explicitly in a global context when the risk information is given to the public. If it is not possible, an individual care approach is more promising of rich implications for the future. Media people must be competent to understand and report the risk information properly with conscience. They must be able to distinguish fact and informed opinion from speculation as it is not easy for the public to understand such explanation in the context of their risk and safety. Massive written articles and snowballing massive explanations on each data set in the Internet produced the status "chaos begets further chaos". To reduce unnecessary stress/fear and untold suffering due to health risks for workers, experts must work together to explicitly explain the complex procedures of the ongoing problems—what we know, what we do not know, what we expect to happen, and what we need to take care of based on timely and proper compilations of each set of raw data from a viewpoint of health protection. To enable spokespeople to provide essential raw data that will reach the population in a timely and proper manner, key points must be noticed and kept in mind. We again and again need to avoid "chaos begets further chaos." In any case, openess of data is the key in the reporting.

Global risk managements at the triple loop learning is required. Contamination of the sea by release of radioactive materials implies very sensitive, ethical and international issues and we need careful scientific surveys on the effects to the sea, which should be followed up by international open discussions. As for our long-term future, we need to prepare reference data for making decisions and organize global discussions open for public. We are now learning a big social risk of nuclear technology, and economic risks and environmental risks are waiting. Proliferation risks exist every time. If nuclear power spreads to less developed countries without deep repositories of expertise, we should consider a regulatory role for the International Atomic Energy Authority, instead of its current role of helping countries upgrade their safety and prepare for emergencies.

Potential natural disasters for each region are definitely to be taken into account now, and all the paradigms that have driven our modern society should be critically reviewed from scratch by global collaborations. We need to consider also risks of other energy sources. The truth is that all energy choices carry risks, and serious and continuous evaluations are required. Climate changes due to greenhouse gas emissions may be more invisible, uncontrollable and complex. Sharing reliable data and having productive global discussions transparent to the public are prerequisites to ensuring our future. More and more transparency and objective reasoning on environmental issues must be organized.

It seems that continuous individual care and reports on health evaluation are the only promising way to verify "safety". So, data-centric approaches with transparent evaluation procedures are the key to overcome the chaos, which is suggesting a necessity for new paradigm on our society as well as sciences. It is a challenge to make our society thoughtful and tough.

8.4 International Collaboration from the Fukushima Accident

Big Earthquake, Big Tsunami and Big Accident, which may happen once in a millennium, have forced us to reconfirm the weakness of our modern society mainly supported by oil and also the weakness of our technology before the nature. Our history to apply mechanics is just 500 years, our history to apply the thermodynamics to learn the earth and technology "cooling based on theory" is just 200 years in age, and our history to use nuclear energy is just half a century.

Valuable lessons learned during and after the TMI and Chernobyl accidents should be reevaluated in order to prepare a set of remedies for Fukushima with different time scales and different radiation exposures, in particular paying attention to socio-psychological aspects. We now have a variety of information channels and contents that must be used to maximum advantage. The dissemination of data and knowledge is far faster and wider than when TMI and Chernobyl occurred, which has various unexpected side effects as well as benefits. The accuracy, timing, and context of the disseminated data must be observed carefully, and advice for timely improvements will become lessons for the future. What general public and active people can add is an analysis of reaction of people to disaster and behavior in critical situation. Reaction and behavior of Japanese people and international friendships and sympathy on disaster of 11 March and Fukushima accidents are to be studied and lessons used around world as an example of peoples behavior. Supports from socio-psychologists are necessary. And it is important to observe that a new feature of the society where we have increased number of sensors, increased number of data sources, meta-language/knowledge/infrastructure for data sharing and exchange, evaluation of data reliability in an era of real-time and near real-time observation and reporting. Diffusion of data, quality of data, semantics of data and contexts of data are interacting each other far more dynamically in our information society that makes changes in public behavior and perceptions.

We are now observing data-driven-real-time-immediate effects on people in emergency situations, which requires us a very careful and deep consideration on its path for the future. A part of the tragedy is some of the media reports, which are somewhat unnecessarily and essentially articulate the fears of the "unknown" and "unexpected". However, scientific analysis is to be based on reliable data and proactive explanations.

Worst-case scenarios are better prepared by the off-site independent experts who have time to think and are calm enough to get a holistic view. They can examine potential interactions between risks, nonlinear responses and feedbacks, common mode failures, and response strategies during extreme events. If in situ data are prepared for open access, we can expect more useful contributions.

8.5 Preparation for "Beyond Fukushima"

Our history to utilize nuclear energy is relatively short, and we need to reevaluate the nuclear technology intensively from scratch. After a short excerpt in 1939 "Über den Nachweis und das Verhalten der bei der Bestrahlung des Urans mittels Neutronen entstehenden Erdalkalimetalle" by Otto Hahn and Frits Strassmann, the era of "nuclear" has evolved making not only good stories but also bad stories. Japanese network of power systems has been developed by introducing nuclear reactors and created a culture of large socio-technical systems and an engine to develop an advanced industrialized society of Japan. We have pursued, in general, a strategy of building larger, more efficient, centralized powerful systems of a minimum redundancy in our modernized society. This Japanese society design had been regarded as exemplars before 3.11 of 2011, but it should be reevaluated at the triple loop learning by Japanese society.

We have been learning many lessons from this accident. Through such trial and errors, decontamination and decommissioning technologies are under developments by getting big intellectual supports from the world. A snapshot of working platform toward this direction is shown in Fig. 8.3. Human dimensions can be crystallized in accordance with the growth of debategraphs of different views guided by an ontology as visualized by colors and arrows.

"Processes are started whose outcome is unpredictable, so that uncertainty rather than frailty becomes the decisive character of human affairs" Hannah Arendt said. To deal with risks, uncertainties, ambiguities, ignorance indeterminacy due to nuclear accidents and climate changes, transparency and data sharing between industry, government, academia, and nongovernmental stakeholders are the keys for productive discussions, which will help in the long run to prevent future catastrophe. It will evoke something productive through international collaborations as described above at different loops, namely, science intensive convergence at the single loop, engineering intensive convergence at the double loop and humanities intensive conversion at the triple loop.

In case of Fukushima Daiichi nuclear accidents, cognitive gaps in evolving features of science and engineering, and locked-in explanations on technologies mattered, which evokes serious needs to carry out reengineering of complicated engineering products after the traceability of associated data and models, and

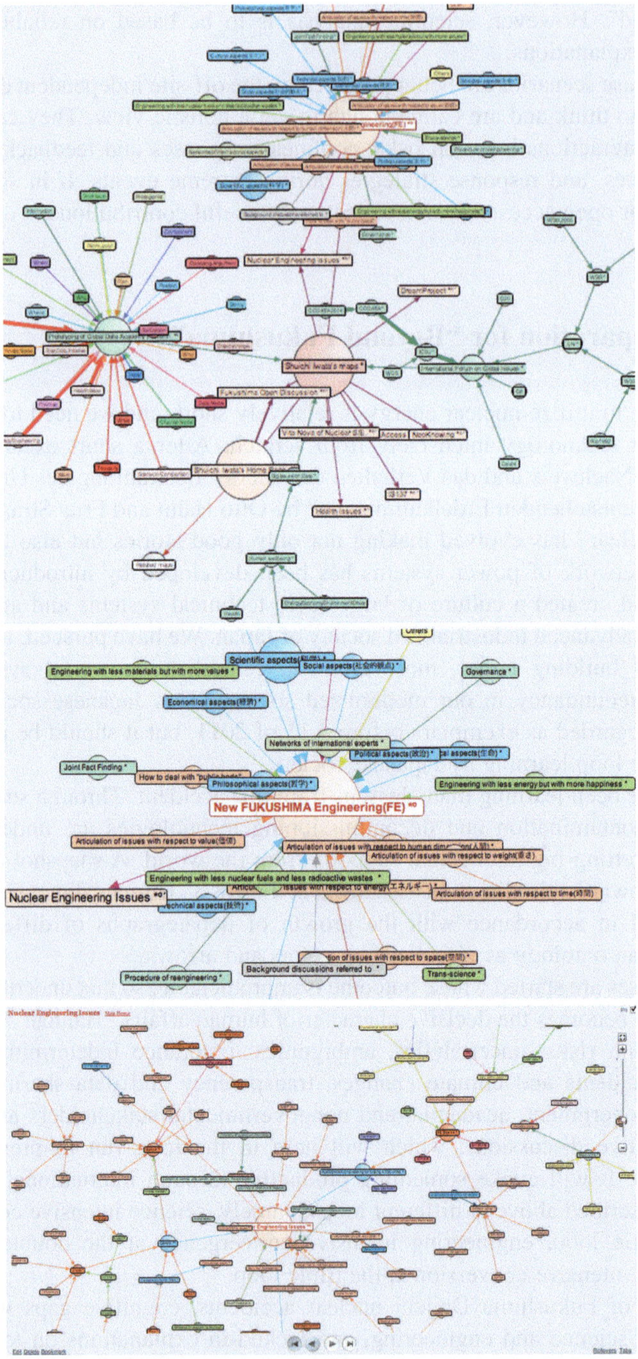

Fig. 8.3 A platform to link nuclear issues of supposing, opposing, optional, task and so on

standards transparent to stakeholders for a consensus. This direction of reengineering are required for all engineering products of unknowns including such tacit knowledge as safety parameters, engineering decisions, know-how, theories and so on, which can be described into a holistic explanation through high quality data linking associated loops simultaneously.

8.6 Concluding Remarks

After the 2011 Tohoku earthquake and tsunami, the accident at TEPCO Fukushima Daiichi Nuclear Plant site happened because the four reactors were not designed to withstand such natural disasters. Reengineering must be done in the design of such nuclear reactors by changing the reference safety rules at the double loop learning level rather than at the single loop learning of the original design. Release of radioactive materials into the environment due to the Accident has transformed the design process to include a much larger question, "Do we need nuclear energy?" at the triple loop learning level. Many interested people and stakeholders with different ideas are shifting from the single loop to the triple loop level of thinking and design. At this time rigorous fact-finding by committed people and stakeholders based on shared data is required to move towards a design solution at the triple loop learning level. That triple loop learning will be enhanced by learning more about facts that relevant experts can handle unexperienced new problems at the Accident site – at the single and double loop levels of learning. Beyond such efforts we can hope that data democracy help us outline solutions for unknowns at new frontiers of science and technology.

Acknowledgement I should like to take this opportunity to express my gratitude to my colleagues, Dr. John McCarthy, Dr. Pierre Villars, Prof. Ying Chen, Dr. Steve LeClair, Dr. Jack Park, Dr. David Price, Dr. John Rumble, Dr. Krishan Lal, Prof. Jun Ni, Prof. Binglin Gu, Prof. Chen Nanxian, Dr. Tony Hey and other colleagues who have pushed me to start a long pilgrim journey of data-driven design science.

References

Chen Y et al (1996) Structural stability of atomic environment types in AB intermetallic compounds. Model Simul Mater Sci Eng 4:1–14

Debategraph (2014) The learning. http://debategraph.org/Details.aspx?nID=250157&lan=EN. Accessed 31 Dec 2013

Ni J (2014) Principles of physics-from quantum field theory to classical mechanics. World Scientific, Singapore

Villars P et al (2008) A new approach to describe elemental-property parameters. Chem Met Alloys 1:1–23

Chapter 9
'Design-Society' Cycle: A Case Study on the Story of Longitude

Amaresh Chakrabarti

Abstract The motive of design should be expanded beyond its current focus on the design process, and should include understanding and possibly improvement of pre- and post-design activities. This chapter undertakes a preliminary enquiry into the broad processes that might constitute design, pre-design and post-design phases, and proposes the 'design-society' cycle as a framework for further enquiry into these processes, to understand their influence on: developing knowledge and experience triggered in the societal mind by a product, subsequent transformation of this knowledge and experience to form new product requirements, and further transformation of this knowledge and requirements to form new products. Experience of individuals and valuation of experience are seen as keys to the development of this knowledge, with the background of the individual, in particular her myriad identities, as a major influence in forming this experience. Using 'the story of longitude' as a case study, the chapter deconstructs the design-society cycle to propose a number of key questions that could form an agenda for further enquiry into the Cycle.

Keywords Collective experience • Design-society interaction • Societal mind

9.1 Introduction

According to Taura (2013), the motive of design should be expanded beyond its current focus on requirements which initiate the design process that spans from initial requirements to a detailed product with the intent to satisfy these requirements (Pahl and Beitz 2002). The expansion should encompass understanding, and preferably improvement, of pre-design and post-design activities, where pre-design involves development of initial requirements from information available in the society, while post-design involves formation of knowledge from experiencing the product in its life cycle.

This chapter undertakes a preliminary enquiry into the broad processes that might constitute design, pre-design and post-design phases, and proposes the

A. Chakrabarti (✉)
Indian Institute of Science, Bangalore, India
e-mail: ac123@cpdm.iisc.ernet.in

© Springer Japan 2015

T. Taura (ed.), *Principia Designae – Pre-Design, Design, and Post-Design*,
DOI 10.1007/978-4-431-54403-6_9

'design-society' cycle as a framework for further enquiry into these processes, to understand their influence on:

- developing the soup of knowledge and experience triggered in the societal mind by a product
- subsequent transformation of this knowledge and experience to form new product requirements, and
- further transformation of this knowledge, experience and requirements to form new products.

Experience of individuals and valuation of experience are seen as keys to the development of this knowledge, with the background of the individual, including her myriad identities, as a major influence in forming this experience. The chapter deconstructs the design-society cycle to propose a series of more detailed questions that could form an agenda for further enquiry into the triad formed by design, pre-design and post-design phases of the Design-Society Cycle.

9.2 Terms

We define several terms for the discussion to follow.

An *experience* is an impression created by an event undergone by an individual. An experience has two components: a description of the event and its context, and its effect on the individual, i.e. how it is valued. For instance, an individual may have visited a medieval festival in Landshut—a medieval town in Germany where its inhabitants enacted the festivities from a wedding that took place in this town in 1475 (description), and found this to be a unique, refreshing and enjoyable experience (value).

Social norms are rules expected to be followed by individuals in a society or societal fragment. For instance, one is expected not to belch in public in many societies in Europe, but is expected to blow her nose in public, even while taking food in a group. The opposite is true in India: especially in the context of taking food, one is expected to belch, but not to blow nose in public. Social norms can have various levels of value, e.g. as to how strongly one is punished for not abiding by these. Some of these norms are enshrined in the laws of the society, e.g. right to one's own life.

Social motives are aspirations of a *societal fragment*—i.e. a collection of individuals within a society). Many social motives are enshrined in incentives and resources for those who buy into the motive. For instance, to increase population can be a social motive in Europe (as can be to decrease population, or provide education for all, in India), and special incentives are devised to encourage each.

Requirements are statements about expected or aspired levels of experience (in the context of a product or solution), and their perceived value for an individual or group. For instance, a requirement for providing 'education for all' might be to develop affordable teaching aids for schools in villages in India.

Expectations are aspects of the world around an individual to which the individual has been accustomed, and the aspects which she wants at least to be retained. Expectations can become requirements if these are unlikely to be fulfilled given the knowledge of reality, e.g. expectations may be to retain the current level of GHG emissions, but given the knowledge of reality where fossil fuel consumption is on rapid increase, this is unlikely to be fulfilled, and may have to be addressed as a requirement.

Aspirations are what an individual wishes the world around her to become. This is typically beyond expectations, where expectations often provide the benchmark against which the intended change will be assessed. A change can be between two different experiences of the same individual, one experience and another from knowledge, one borrowed from the aspirations of valued others, a change between one's experience and that from a social norm, a borrowed social motive, and so on. For instance, one may be used to garbage being handled in an unsafe manner by people from low income groups (own experience), and aspire the situation to change where garbage will be handled in a safe manner by people who would be well paid for the job (e.g. from knowledge of other contexts), etc.

Identity of an individual is a label signifying a group to which the individual perceives herself to belong, or to whom others perceive the individual to belong. For instance, an individual may be perceived to be an Indian by nationality, a Hindu by birth, a male by gender, an engineer and a PhD by education, a professor by profession, a husband, a father or a member of a specific family, a Bengali by mother tongue, etc. Each identity provides a rich tapestry of experiences, and a variety of people with whom the individual has the chance to share or juxtapose her experiences and their values.

9.3 How Are Aspirations and Expectations Created?

Chakrabarti (2013) proposed five major factors that influence an individual's experience (i.e. action and learning), leading to their subsequent modification: attitude and interest, resources, incentives, knowledge, and remunerations. These are influenced by environment via incentives (goals, promise for remunerations) and resources (physical resources and external knowledge) in the beginning, and external remunerations at the end of the action and associated learning event. The question we ask now is: what is responsible for the construction of expectations and aspirations? Expectations, we propose, are influenced by (Fig. 9.1):

- Experience of the individual, which is part of the knowledge formed as the individual undergoes the actions and learning in her various identities, and how these are valued by the individual;

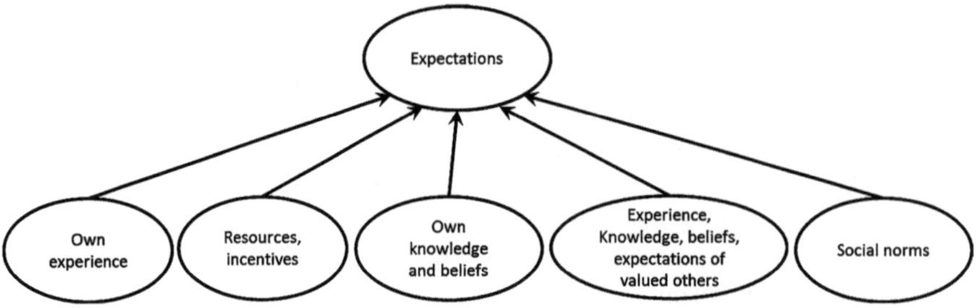

Fig. 9.1 Possible factors responsible for construction of expectations

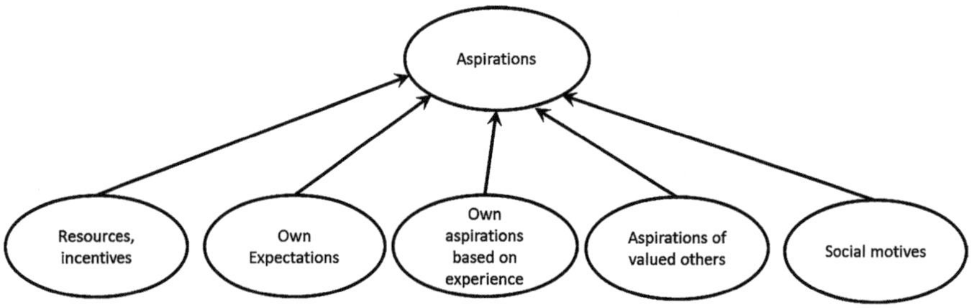

Fig. 9.2 Major factors responsible for construction of aspirations

- Experience, knowledge, beliefs and expectations of other individuals who are valued by this individual; This is part of the incentives provided by the environment;
- Social norms around the individual and how the individual values these norms; this is part of the resources and incentives provided by the environment;
- Knowledge and beliefs of the individual: this contains knowledge about similar or analogous experiences and their values, which come from sources other than the three points mentioned above.

Aspirations are based on various factors: The primary ones are (Fig. 9.2):

- Expectations of the individual and associated factors; this is part of the interest and attitude of the individual;
- Aspirations of the individual based on her own experience; this is part of her interest and attitude;
- Aspirations of those who are valued by this individual; this is part of the incentives from environment;
- Social motives around the individual and how the individual values these motives; this is part of the incentives provided by the environment;
- Resources and incentives provided by the individual or the environment.

9.4 Transformation of Individual Expectations and Aspirations into Social Norms and Motives

We propose the following processes to be primarily responsible for the transformation of individual expectations and aspirations into social norms and motives (Fig. 9.3):

- Transformation of individual expectations and aspirations into collective expectations and aspirations: Similar experience of many individuals reinforces its strength, often with canonization. The value of the experience is increased and often institutionalized through this collectivisation process, e.g. forming organizations for seeking justice for the Bhopal Gas Disaster victims (http://en.wikipedia.org/wiki/Bhopal_disaster).
- Transformation of collective expectations and aspirations into societal norms and motives: Collective expectations get transformed into societal norms, e.g. laws, cultural norms, etc. Collective aspirations get transformed into social motives.
- De-contextualization and re-contextualization of norms and aspirations: Norms and aspirations often get detached from the original experience, and become part of the cultural elements of a societal fragment. De/re-contextualization is a key aspect of how a society forgets, merges or changes the context, e.g. the problematic experiences which a norm or expectation had tried to overcome, thereby providing both barriers and opportunities for bringing in new changes.

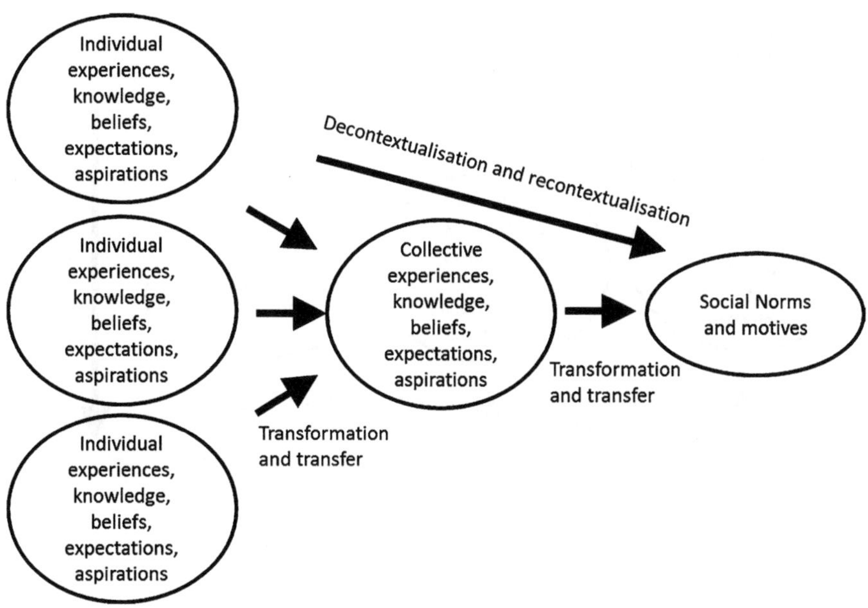

Fig. 9.3 Construction of collective experiences, expectations, aspirations, social norms, motives

De/re-contextualization may happen as individual experiences, expectations, or aspirations become collective (e.g. as similar or analogous experiences are constructed out of these), or are further transformed into social norms and motives.

9.5 How Are Requirements Created?

Requirements for change are created by an individual, a team, a community or an organization, by transforming, transferring, decontextualizing and/or re-contextualizing the following (Fig. 9.4):

- Experiences, expectations or aspirations of an individual: The individual can be the designer herself, or someone else. The expectations or aspirations can pertain to any part of the lifecycle of the proposal.
- Collection of experiences, expectations or aspirations of other individuals: This is a set of experiences, expectations or aspirations collected from multiple individuals, which are yet to coalesce into a 'collective' experience as in the next bullet point, where 'collective' refers to experiences, expectations or aspirations and their context that are held as similar by a collection of individuals.

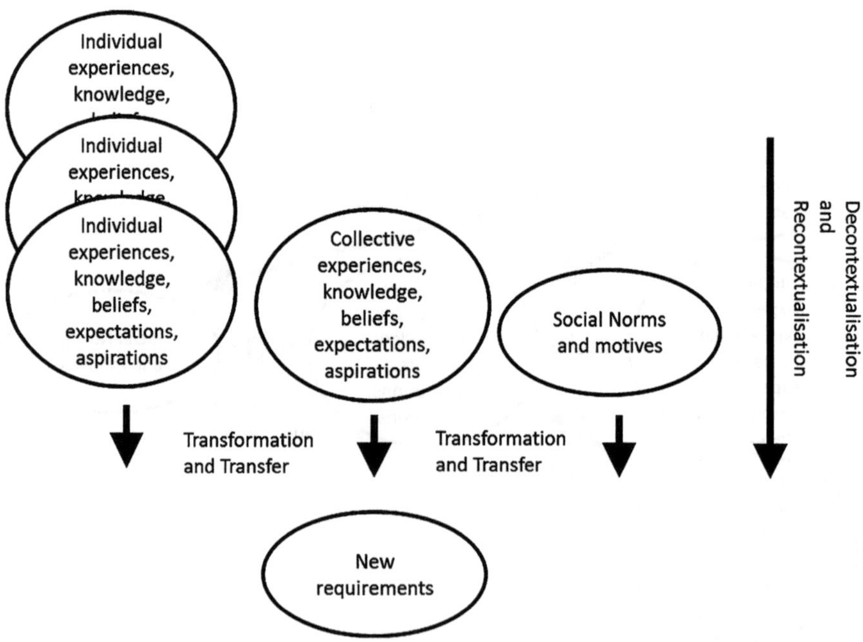

Fig. 9.4 Construction of new requirements

- Collective experiences, expectations and aspirations. The collective can be a community that is bound by a combination of identities, and have aligned experiences and contexts. An example of collective aspiration might e.g. be that of young, educated people from a community in India to become officers in the Indian Administrative Service. An example of collective aspiration, for victims of Bhopal Gas Disaster (http://en.wikipedia.org/wiki/Bhopal_disaster), is to get adequately compensated for by the US multinational Union Carbide.
- The norms and aspirations of a societal fragment.
- A combination of these.

9.6 Design-Society Cycle

The design-society cycle, as a combination of design, pre- and post-design phases, is proposed in Fig. 9.5. During pre-design, requirements are created by processes of transfer (acceptance without change in content), transformation (acceptance after change in content), de-contextualization (acceptance after removal of context), or re-contextualization (acceptance after changing the context) of experiences, expectations, and aspirations of an individual, a set of individuals or a collective, and social norms and motives. These constitute the main pre-design processes.

Starting with new requirements, new products (the term is used in the most generic sense possible, to include all enablers of change) are developed. The activity requires processing of experiences, beliefs, knowledge, expectations, and aspirations of individuals and collectives, and of social norms and motives.

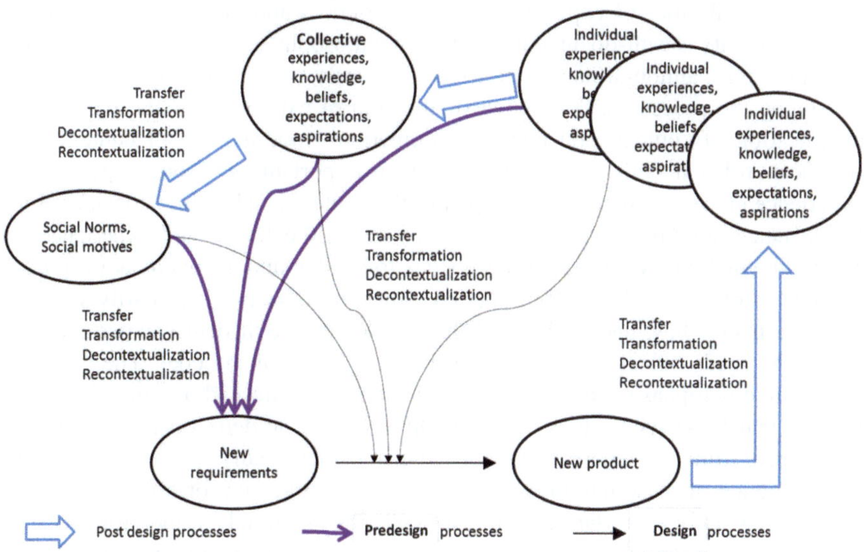

Fig. 9.5 Design-society cycle

The new product in its lifecycle variously influences experiences, and as a result expectations and aspirations of individuals. These transfer-, transformation-, and de/re-contextualization-processes enable modification of the experiences, expectations and aspirations of the collective, eventually modifying social norms and motives. Each change operates at a different time scale: formation of individual experience happens faster than formation of collective expectations, which is faster than formation of social norms.

9.7 Role of the Various Processes

In the above model, experience and its valuation play a major role. We argue that the identities of the individual play a significant role in the processes applied in the formation of the individual's experience, which, as conjectured before, has a major influence on the construction of expectations and aspirations.

There are various kinds of identity that are held by, or impressed upon individuals. Social sciences speak of five major social institutions: family, education, religion, politics and economy (http://sparkcharts.sparknotes.com/gensci/sociology/section9.php). Each provides a collection of identities; as a result of its unique history, expectations and aspirations, each such identity provides both opportunities and barriers to experiencing reality and valuing such experiences. Since identity is unique for each individual (even within the same family, and particularly since each individual traverses a different trajectory of identities in her life), the same exposure to reality could evoke a different experience and its valuation. Further, each identity exposes one to different sets of experiences, expectations and aspirations, leading to different possibilities of exposure to new experiences and their valuation. Importance of identity is well-documented in Indian philosophy (e.g. http://www.iep.utm.edu/hindu-ph/), although its role in design has rarely been explored.

Among the processes, de/re-contextualization of experiences—the removal or modification of context, and their transformation—modification of content play significant roles. In both, analogical reasoning is important: only some aspects of an experience are remembered, or only part of the context is taken into consideration. For instance, an Austrian engineer took a walk in the forest, and experienced a reality in which many cocklebur seeds stuck to his dog and his overcoat (http://en.wikipedia.org/wiki/Velcro). The experience must have been at least partly unpleasant, in terms of 'how badly' the seeds had stuck, and the difficulty experienced in taking the seeds out of the hair of the dog and fabric of the coat. The engineer seemed to de-emphasise this part of his experience, and instead became excited by the fact that these seeds stuck 'so well'. In doing so, he de/re-contextualized the experience, from one where cocklebur seeds got stuck 'badly' to 'well', and from an experience of these natural seeds sticking to his overcoat or dog to a goal of attaching something similar but artificial to specified materials. It therefore became an aspiration for the individual, which eventually formed a requirement that led to the invention of 'velcro'.

9.8 The Story of Longitude and John Harrison

John Harrison (1693–1776) was an English inventor and horologist, who solved one of the most difficult problems of the eighteenth century: how to determine the longitude of a ship at sea, saving many lives (https://www.asme.org/engineering-topics/articles/mechanisms-systems-devices/john-harrison). Harrison was fascinated with watches, clocks, and other timepieces since the age of six when he was sick with smallpox, and had to entertain himself with a watch his parents placed on his pillow. At that time, watches were large, and it was easy to see how they worked. Harrison's father was a carpenter, but also repaired clocks; Harrison started assisting his father in his work as soon as he was old enough. Further on, he combined his interest in woodworking and timepieces and started building clocks. At 1713, at the age of 20, he completed his first grandfather clock. A year later, Parliament offered a prize of 20,000 pounds to calculate a ship's precise longitude at sea. Harrison decided to compete for this prize.

The existing principle of calculating longitude was the following. For every 15° travelled westward, the local time moved back by an hour. If the local time and time at another known point on the Earth were known, the difference could be used to calculate longitude. While the local time could be estimated by observing the sun, no reference point such as GMT was available for calculation of the other time. The only clocks then were pendulum clocks, which became inaccurate by a ship's motion and temperature changes.

Determining longitude on land was relatively easy, due to a "...stable surface to work from, a comfortable location to live in while performing the work and the ability to repeat determinations over time made for great accuracy" (http://en.wikipedia.org/wiki/History_of_longitude). For longitude, early ocean navigators had to rely on 'dead reckoning'—calculating one's current position by using a previously determined position, and advancing that position based on known or estimated speeds over elapsed time, and course (http://en.wikipedia.org/wiki/Dead_reckoning). This was inaccurate on long voyages out of sight of land.

To avoid problems with not knowing one's position accurately, navigators sometimes relied on taking advantage of their knowledge of latitude, which was relatively easy to calculate from declination of the sun. Sailing to the latitude of their destination, they would turn toward their destination and follow a line of constant latitude (Dunlap and Shufeldt 1972). However, this solution prevented a ship from taking the most, or most favourable, route, extending the duration of the voyage and increasing the risk of poor health or death to members of the crew, and the resultant risk to the ship (http://en.wikipedia.org/wiki/History_of_longitude).

Navigation errors often resulted in shipwrecks. Motivated by a number of such maritime disasters attributable to serious errors in reckoning position at sea, e.g. the Scilly naval disaster of 1707 where a British naval fleet, along with its Admiral Sir Cloudesley Shovell, sunk off the Isles of Scilly, the British government established the Board of Longitude in 1714 (http://en.wikipedia.org/wiki/Scilly_naval_disaster_of_1707): "The Discovery of the Longitude is of such Consequence to

Great Britain for the safety of the Navy and Merchant Ships as well as for the improvement of Trade that for want thereof many Ships have been retarded in their voyages, and many lost..." [and there will be a Longitude Prize] "...for such person or persons as shall discover the Longitude."

The prizes were planned to be awarded for the demonstration of a practical method for determining the longitude of a ship at sea, in various amounts for solutions of various accuracy. These prizes, worth millions of dollars in today's currency, motivated many to search for a solution (http://en.wikipedia.org/wiki/History_of_longitude).

Besides Britain, France, Holland, and Spain were also interested in solving this problem. France's King Louis XIV founded Académie Royale des Sciences in 1666, with an aim, among others, advancement of the science of navigation and the improvement of maps and sailing charts. From 1715, the Académie offered a prize specifically for navigation (Taylor 1971). Spain's Philip II offered a prize for the discovery of a solution to the problem in 1567; Philip III increased the amount of the prize in 1598. Holland offered a similar prize in 1636 (http://www-groups.dcs.st-and.ac.uk/~history/PrintHT/Longitude1.html). Navigators and scientists in most European countries were aware of the problem and tried to find a solution, making it one of the largest scientific endeavours in history.

Astronomers, however, believed that the solution lied in mapping objects in the sky. The first publication of a method for determining time by observing the position of the Earth's moon was in 1514 by Johannes Werner (http://en.wikipedia.org/wiki/History_of_longitude); other proposals followed from Galileo, Hally, Maskelyne, etc., although with little success as the proposals were unreliable due to errors introduced by the sea or variability in the phenomena used.

Harrison, however, looked for a mechanical answer to meet the strict criteria of keeping a ship within half a degree of longitude on a voyage from England to the West Indies. For this, a clock had to maintain an accuracy of 2.8 s a day, a huge challenge against temperature changes and a ship's motion.

The concept of using a clock was probably introduced in 1553 by Gemma Frisius (http://en.wikipedia.org/wiki/Gemma_Frisius). Pendulum clocks had been used on land with success; Huygens had developed accurate pendulum clocks that were used to determine longitude on land. He also proposed the use of a balance spring to regulate clocks. Many, including Isaac Newton, however, thought it impossible to develop a clock of the required accuracy.

Harrison began working on a land clock that was more accurate than any of the time. He revolutionized one aspect of the clock by eliminating the need for lubricating oil that often contributed to clock-failure. Oils dried out in summer and became thick in winter. Working with his younger brother James, in 1720 he designed and built a clock-tower for the family-estate of Sir Charles Pelham. With this clock, he made further design changes that enhanced the stability of the clock. They designed a series of high-precision clocks that included innovations such as a pendulum rod made of alternate wires of brass and steel, eliminating the problem of a pendulum's effective length varying with changes in temperature.

With confidence from the performance of his land clocks, Harrison began working on a sea clock, and over a period of 40 years produced a series of timekeepers, now referred to as H1, H2, H3, H4 and H5. His first, H-1, was not tested under the conditions required by the Board of Longitude. Instead, the Admiralty wanted it to travel to Lisbon and back. It performed excellently, and the Board awarded Harrison a grant of 500 Pounds for his work. However, the "perfectionist in Harrison prevented him from sending H1 on the required trial to the West Indies" (http://en.wikipedia.org/wiki/Dead_reckoning). Instead, he embarked on the construction of H-2, which never went to sea, and was immediately followed by H-3. H-3 incorporated two design innovations: a bimetallic strip to compensate the balance spring for the effects of changes in temperature, and a caged roller bearing, the best of his anti-friction devices (http://www.rmg.co.uk/harrison). However, neither H2 nor H3 was accurate enough, so Harrison went on to produce H-4. While the first three were large clocks with special balance mechanisms compensating for the ship's motion, H4 was a timepiece that resembled a large pocket watch, and greatly exceeded the requirements for the prize.

However, for various reasons including that some of the members themselves wanted to win the prize, the Royal Society awarded him only a portion of the money and asked for more tests. Nevil Maskelyne, the then Astronomer Royal, remained unconvinced that a watch could be more reliable than the lunar distance method for finding GMT (http://www.rmg.co.uk/harrison). Meanwhile, an expedition by Captain Cook with a copy of H4 proved beyond doubt that longitude could be measured from a watch (http://www.rmg.co.uk/harrison). It eventually took the intervention of King George III to get Harrison, who by then was 80 years old, his full reward and recognition, some 12 years after he had fulfilled the original conditions.

Though the British Parliament rewarded John Harrison for his marine chronometer in 1773, his chronometers were yet to become standard. Chronometers such as those by Thomas Earnshaw were suitable for general nautical use by the end of the eighteenth century. However, they remained expensive and the lunar distance method continued to be used for some more decades. Eventually, by 1850, most of the ships worldwide had stopped using the lunar distance method (http://en.wikipedia.org/wiki/History_of_longitude).

9.9 Design-Society Cycle in the Story of Longitude

The story of development of chronometers by John Harrison demonstrates the importance of experience, beliefs, resources and incentives in the development of expectations and aspirations of an individual. His experience with clocks at early stages of his childhood was instrumental in fuelling his fascination for timepieces; his experience of assisting his father in clock-repair was valuable in forming early expectations and aspirations with the capability of clocks. His experimentations with the accuracy of land-clocks led to innovations in anti-friction devices and stabilizing mechanisms, which fuelled new expectations and aspirations in developing clocks of further accuracy. Further, the knowledge that longitude on land was possible to determine with accuracy (e.g. by Huygens) gave further strength to this aspiration of building such clocks for the sea—an example of formation of aspiration as de/re-contextualization of experience, knowledge or beliefs—in this case from accurately determining longitude on land to that on sea.

On the other hand, the majority in the rest of the world had a very different experience and knowledge of clocks. Their experience with clocks of lower accuracy and their knowledge of the failure of pendulum clocks at sea, combined with their (e.g. for Maskelyne) strong belief that the answer lied in using astronomical phenomena, led them to form very different expectations and aspirations.

Multiple shipwrecks, including major ones such as the Scilly disaster, led to formation of collective experience in the UK, which associated the current solutions—'westing' or 'dead reckoning' as, respectively, with long-and-wasteful or error-prone-and-high-risk voyages. This collective experience eventually led to formation of a social motive of developing an accurate solution to the problem, which was institutionalized in the form of the Board of longitude and in the Longitude prize. The longitude prize, worth millions of dollars in today's currency, provided a strong incentive; further resources were also provided by the Board as grants. These created aspiration in many to search for a solution.

As can be seen, the pre-design processes consisted of experiences with current solutions for sea- and land-navigation, which led to collective experience, beliefs and knowledge about these solutions, with the eventual formation of similar social motives across societal fragments (Britain, France, Holland...), and their institutionalization in terms of resources and incentives (board, grants and prizes, etc.). All of these formed the backdrop and influences for further pre-design processes, which took very different paths in different individuals. Newton thought it impossible to achieve the task. Maskelyne, with his firm beliefs in Astronomy and earlier experiments with astronomical observations as sea, believed the answer lied only in astronomical phenomena. Harrison, on the other hand, with his knowledge and experience with timepieces believed the answer lied in chronometers, and spent 40 years developing such a solution.

The design-society cycle is rarely concluded; with all their accuracy, even H4 and H5 were too expensive to be accepted as a common solution. The lunar distance method continued to be used for decades after, as new pre-design efforts to develop more cost-effective timepieces began continued in parallel.

9.10 Issues for Further Research

The case study presented in this chapter is meant to stimulate discussion into possible ways of investigating the interactions between design and society, for taking into account, or even influencing the way pre-design, design and post-design processes (should) interact. The proposed 'Design-Society' Cycle is one possible framework for inquiry into such interactions.

A number of questions emerge from the discussion. Some are listed below. The first set of questions is related to post-design processes:

- What are and should be the processes involved during post-design?
  - How are transfer, transformation and de/re-contextualization carried out in post-design?
  - How (well) are these processes carried out in the formation of experience, expectations and aspirations and their values?
  - How (well) are these processes carried out in the formation of experience, expectations and aspirations of the collective?
  - How (well) are these processes carried out in the formation of social norms and motives?
  - How should transfer, transformation and de/re-contextualization be carried out in the post-design processes for improvement of pre-design and design processes?
- What is the role of identity in the construction of experience and its valuation?
- What is influential in the formation of expectations?
  - What is the role of experience?
  - What is the role of knowledge?
  - What is the role of expectations of others valued by the individual?
  - What is the role of social norms?
- What is influential in the formation of aspirations?
  - What is the role of own expectations?
  - What is the role of aspirations of others valued by the individual?
  - What is the role of aspirations based on own experience?
- What is the role of social motives?

The second set of questions is related to pre-design processes:

- What are and should be the processes involved during pre-design?
  - How are transfer, transformation and de/re-contextualization carried out in pre-design?
  - How (well) are these processes carried out in the formation requirements and their values from experience, expectations and aspirations of individuals?
  - How (well) are these processes carried out in the formation of requirements from experience, expectations and aspirations of the collective?
  - How (well) are these processes carried out in the formation of requirements from social norms and motives?
  - How should transfer, transformation and de/re-contextualization be carried out in the pre-design processes for improvement of post-design and design processes?
- What is the role of identity in the construction of requirements and their values?

The third set of questions is related to design processes:

- What are and should be the processes involved during design?
  - How are transfer, transformation and de/re-contextualization carried out in design?
  - How (well) are these processes carried out in the development of requirements, product and their values from experience, expectations and aspirations of individuals?
  - How (well) are these processes carried out in the development of requirements, product and their values from experience, expectations and aspirations of the collective?
  - How (well) are these processes carried out in the development of requirements, product and their values from social norms and motives?
  - How should transfer, transformation and de/re-contextualization be carried out in design processes for improvement of pre-design and post-design processes?
- What is the role of identity in the development of requirements, product and their values?

One potential issue is the difficulty of carrying out empirical research into the design-society cycle. For instance, effort required to unearth historical detail required might be huge, and might require design researchers to work together with historians. Further, the number of events in a cycle can be too many to be investigated in depth. A possible alternative might be 'to focus only on significant events' as a method. For instance, the story of Longitude seems to have only a handful of significant events: Scilly disaster, formation of Longitude Prize, experience of Harrison with clocks, experience of Maskelyne on astronomical phenomena, conflict of interest between aspiration and duty of Maskelyne, knowledge of previous solutions and their value, and intervention of the king. This raises further methodological questions: which events are significant, and how should significance be decided so that research can be carried out with external validity?

References

Chakrabarti A (2013) Understanding influences on engineering creativity and innovation: a biographical study of twelve outstanding engineering designers and innovators. Int J Des Creat Innov 1(1):56–68

Dunlap GD, Shufeldt HH (eds) (1972) Dutton's navigation and piloting, 12th edn. Naval Institute Press, ISBN 0-87021-163-3

Pahl G, Beitz W (2002) Engineering design – a systematic approach. Springer, London

Taura T (2013) Motive of design: roles of pre- and post-design in highly advanced products. In: Chakrabarti A, Blessing LTM (eds) An anthology of theories and models of design: philosophy, approaches and empirical explorations. Springer, London

Taylor EGR (1971) The Haven-finding art: a history of navigation from Odysseus to captain cook. Hollis & Carter, London. ISBN 0-370-01347-6

Chapter 10
Negative Technology: Possibilities for a Contribution from Eco-Ethica—A Combination of Ethics and the Invisible

Noriko Hashimoto

Abstract The characteristic feature of the end of the twentieth century is a modernized, systematized society for the development of technology. By preserving the character of instrumentality, technological cohesion or conjuncture is our new circumstance. New phenomena have arisen such as technological abstraction, a separation of function from form, and cards with many functions. Until the twentieth century, technologists looked for new technology as a positive way to create artifacts. But in the twenty-first century, it is necessary to think about negative technology (technica negativa). Negative technology is no longer a theoretical idea but a practical project. After Fukushima, we must concern ourselves about safety and close the nuclear facilities. Such a project must be a negation of positive technology, but this negativity should be turned to be positive.

Keywords Abstraction • Eco-ethica • Metatechnica • Negative technology • Technological conjuncture

10.1 Introduction

On 11 March 2011, I experienced the extraordinary earthquake and tsunami in Japan. I have never experienced such a strong earthquake with complicated movements. Even now, I can feel and remember the awful trembling from the depths. When I saw the documentary videos of tsunami, which I could never watch at that time, I was moved and feel a profound sympathy with people in stricken areas. And when in 2013, the disaster in the Philippine islands occurred in the wake of global warming, we Japanese had the earthquake and tsunami in mind. I am sure that disasters we are experiencing these days from the global and climate problems will change our worldview and turn people's thinking upside down. In other words, our view of the world necessarily changed quickly, but it is also fortunate for us to live

N. Hashimoto (✉)
Aoyamagakuin Women's Junior College, Tokyo, Japan
e-mail: hnoriko@luce.aoyama.ac.jp

© Springer Japan 2015

T. Taura (ed.), *Principia Designae – Pre-Design, Design, and Post-Design*,
DOI 10.1007/978-4-431-54403-6_10

together on the one and same globe: symbiosis with environmental circumstances for living together with others. In the twenty-first century, we must conceive an environmental philosophy to account for a triple set of circumstances: namely, (1) nature, (2) technology, and (3) traditional and historical cultural circumstances.

10.1.1 Nature

Nature itself was a friend of humanity in many ways. We can look for the way of symbiosis with nature. For example, the Chinese theory of art has had great influence on Japanese artistic activities. According to the Chinese aesthetician, Cho Gen-en (張彦遠) in his 品等論 (which means the theory of judgment on the value of works of art), the first and most important principle is "to express Chi (気) in nature, a vivid expression of power or energy in natural things". And another aesthetician Wan-wei (王維) said that, in depicting a landscape, painters must express the Chi of the mountain, not necessarily the exact appearance of the mountain. This means that the artist must harmonize with nature. It is different from the European idea of representation or mimesis. But we Japanese recognized that nature sometimes revolts against human beings. Looking at videos or documentary films of the tsunami carefully, we found the approach of waves was so quick and the surface and inner movements of the water crashed and moved in contradictory ways. We felt the immense power or energy in the inner movement of the awesome, disastrous tsunami.

10.1.2 Technology

Gabriel Marcel (1889–1974), wrote that technology as the embodiment of logic and technology itself is originally good (*bien*), but he indicated that, if human beings were to become functions of a systematized technology, they would become subhuman (*sous–humaine*). And he said that modernized technology may easily become an idol and, as the result, make technologist himself an idol (*autolâtrie*), because he could create such splendid technology. But with the nuclear accident following the earthquake and tsunami, we recognized that the forefront of technology could never have been controlled. And the hydrogen explosions in the nuclear facilities in Fukushima occurred contrary to our common sense concerning artifacts. We could not predict what might happen.

10.1.3 Culture

Every people on one and the same planet has its own culture. Culture itself is an important axis related to its past, present, and future. Every culture has its historical tradition and its temporary meaning—and a special space such as its milieu. Every culture is always in many respects in a good relationship with nature (successful harmony with nature and coincidence with nature), and the progressive development of technology makes our life more convenient and happy. The development of technology makes our life fruitful. And the problems in a modern city of our computer-time are the same and common. It is because the computer connects different cultures: we have now an information society. But we must recognize that new conflicts between cultures or the modern city have arisen. The domain of culture is complicated, and we are looking for the possibilities of a reasonable universality.

The characteristic feature at the end of twentieth century is the time for a modernized and systematized society for the development of technology. The Japanese philosopher, Tomonobu Imamichi (1922–2012) named this developed, systematized technology technological cohesion or conjuncture. Technological cohesion or conjuncture means: by preserving the character of instrumentality, newly developed technology is combined with computers and electronic networks and, as a result, technology itself is becoming our circumstance to a large degree. Human attitudes are now controlled by such a circumstance as technological conjuncture or cohesion (cf. Imamichi, vols. 1–24).

10.2 Technological Cohesion or Technological Conjuncture

10.2.1 Abstraction

What is the core problem in the contemporary technology? It is abstraction. Gabriel Marcel said that he always fought against abstraction in general and, in particular, the abstraction of technology. Because he aimed for a concrete philosophy, he discovered the philosophical problem: despair in everyday life in technological world.

What is abstraction? The term abstraction is used only for the logical process of making a concept, by producing one important concept from common elements by eliminating the idiosyncratic. The ontic perceptual abstraction is done by visual and audible sensibility. At the end of nineteenth century, we can find the abstract work of drawing. Herbert Read wrote in his book *Philosophy in Modern Art:* "by abstraction, we mean that it is derived or disengaged from nature, the pure and essential form abstracted the concrete details.—Abstraction will include any form of expression which dispenses with the phenomenal image, and relies on elements of expression that are conceptual, metaphysical, abstruse, and absolute." Abstraction is a higher mental operation that is conscious and, at the same time, unconscious. Observing nature, artists such as Braque and Picasso extracted the essential form from phenomenal things and made a new image through the process of (1) analytic cubism and (2) synthetic cubism. As the result, they succeeded in putting the multi-viewpoints (three-dimensional perspective) into a two-dimensional space. It is a trial for transcending perspective. Through abstraction, we can see a new dimension in the phenomenal world.

Imamichi stressed abstraction in the beginning of modern technological times as following:

> The revolutionary, extensional abstraction has been opened in the last 200 years, that is to say, technological abstraction as energetic abstraction. At first, the human being extracts invisible power from water, coal, oil, etc. The power itself is not visible, but it is supported by heat in the thermodynamic process. Steam, fire, and electricity are visible, but power itself is invisible. In this sense, technological abstraction must be called an energetic abstraction because the object of the abstraction is visible and only the effect is invisible. This extensional invisibility as energy becomes the object of new science as technology and as nuclear science. Recently, namely within the last 60 years, technological abstraction has produced nuclear power, which can be turned into electricity for modern public life and/or can be used for nuclear weapons. That means that the result of the technological coherence exists in both senses, namely, for convenience in the good sense and for destruction on a global scale in the bad sense. (Imamichi 2003, p. 46)

According to Imamichi, these 60 years of technological abstraction created a distinction between "visible" and "invisible". But technological abstraction, founded on logical abstraction, is seen from the side of creative technologists. At the final stage, we users find the most convenient tools easily and pretend to understand the structure. We use smart-phones with thumb/finger to touch the screen or move the screen by two fingers. But no one knows the technological structure as such a convenient tool.

10.2.2 Iso-Morphic and Hetero-Function

What is most important for the machine? In other words: What is the essence of machine? It is function. But, in old days, the idea of the machine came from "form". For example, the telephone: at its beginning, the form of the telephone is composed of the forms of ear and mouth in the idea of speaking with and listening to another person easily: the machine's form was determined by its original functions: loudspeaker signifies a large mouth, and the sensor for hearing signifies an ear's function. By developing the machine's functions, a little black box on the table shows complicated functions. In the old days, a little black box shows the same form, and it was not easy to distinguish between a tape recorder or a camera or a bomb (cf. Imamichi vols. 1–24). As times passed, the developed machine grew smaller and smaller and thinner and thinner: finally, form no longer signifies its proper function. The machine itself is getting small and thin, as it becomes possible to put it in a pocket: everything is to be the form of a thin card. Nowadays, we have many kinds of cards: identity card, credit card, shopping card, a card for opening the door, key card, etc. The most characteristic feature of cards is that they are full of information for identification. It is convenient to use these cards. We can use a card, if I can identify my own number, as invisible money. Finally, we use cards as a core of our ordinary life, or we are dominated by the system of cards: the form of the card is the same thin and easily accessible form (iso-morphic), but each card functions differently (hetero-function). So, at the first sight, we cannot recognize the aim of the machine. It is invisible.

Today, even the personal computer is composed on the form of the iPhone or smart-phone or tablet: very fine but very small and thin. Through the screen, we can contact others and get the news and information from all over the world at a certain moment. This tablet has a dictionary function and contains resources of historical and intellectual knowledge.

10.3 Technological Abstraction and Technological Conjuncture or Cohesion

What is abstraction in technological conjuncture? It is different from logical abstraction.

The ultimate aim of technological conjuncture or cohesion is to diminish the time process, that is to say, to do everything rapidly, thinking only of the effective result. It is possible to combine machines with computers. Economic effects are especially clear and distinct (*clara et distincta*) results of technological cohesion or conjuncture. The technology itself is an embodied system of logics. As a result, process itself became technologically invisible, and we enjoy our life unconsciously by using the keyboard or touching the screen with thumb and fingers to introduce persons into the technological conjuncture or cohesion. We are surrounded by such a circumstance and easily enter into the Internet system.

Thus, the analysis of the technological conjuncture or cohesion is very fruitful, because we find new concepts such as neighbor newly defined insofar as neighbors are connected by technological conjuncture or cohesion by the Internet or iPhone or smart-phone on a worldwide scale. And a new kind of friendship has occurred in a world – namely, a kind of virtue, *philoxenia* (love of strangers) (cf. Kemp, Philoxenia—une vertu eco-éthique, read in Tokyo, 2010, at 29th eco-ethica Symposium). The technological conjuncture or cohesion makes for a new, peaceful relationship between people.

In 1965, Tomonobu Imamichi presented "eco-ethica" as the new ethics for the technological era. 'Eco' means house or, in a wider meaning, a living place or environment. But through technological conjuncture or cohesion, human activities were enlarged, on the one hand, in the direction of cosmic intersidereal space and, on the other hand, in the direction of microscopic nanospace. Eco-ethica is the fundamental philosophical ethics for such a wide space of human action. As a first step, it is necessary to think philosophically about technology itself. But it is not only an objective observation on phenomenal technology but also on the existence of human beings and the ethical attitude to nature, technology, and other human beings.

10.4 Metatechnica and Technica Negativa

10.4.1 What is Metatechnica?

Metatechnica is a new term coined by Imamichi in Eco-ethica to help research new technological problems related to modern circumstances. Greek preposition "meta" means "after or beyond" and "technica" means technology - namely, to think about modern technology. In briefly, "after and beyond" research of technology itself, we must investigate meta-technica as an important problem in the twenty-first century. Like Aristotle's *Metaphysics* is "after and on (beyond)" research into physics through observations of phenomena in nature, Imamichi thought more profoundly concerning the "Being of human existence and God". A new question then occurs: What is the purpose of Metaphysics? Aristotelian thought was teleological and went beyond *ousia* (existence). According to his *Nichomachean Ethics*, his aim is the superior goodness, the best, i.e., happiness. (cf. Imamichi, vols. 1–24).

Based on philosophical reflection suggested by Aristotle, the aim of technological conjuncture or cohesion is happiness. The new question arose: "what is happiness in the contemporary situation, namely, in the era of technological conjuncture or cohesion?" In other words: What is happiness in the twenty-first century? It is "living on the same globe" or living with the others even if they are in an entirely different country or even if they belong to different race, having the same consciousness to live together and knowing each other on the same globe.

Meta-technica is a philosophical reflection on this newly-developed perspective on human existence that has opened for the future of society. And it is necessary to analyze modern technology within the perspective of eco-ethica. The terms Technica Positiva and Technica Negativa were proposed by Imamichi in 1998. (Imamichi, Theoretical Interaction between Metatechnica and Eco-ethica, pp. 1–10. in vol. 16). Technica Positiva includes ordinary human activities for making new products, i.e., artifacts: technology itself is always against natural phenomena – against the dynamic "*Werden*" (becoming of nature), so to speak. Technology builds artifacts vertically on the ground: its positive activities are against the direction of geotropism (growth of a living organism in response to gravity, as the downward growth of plant roots).

10.4.2 Technica Positiva and Technica Negativa

The design for new technology might aim at the effective production of new, convenient, and safe products (artifacts). We human beings want to have richer and more convenient lives, and technologists make creative designs to realize the dreams and desires of people. Nowadays, consumers may also participate in the process of creative activities from the beginning. This means that the definition of consumer has changed. Consumers have become makers of new products (artifacts) along with the producers. In a way, we can say that their desire will be realized through positive activities for making new things. Such activities may be called Technica Positiva. In the old days, people were normally accustomed to making new products, even for their own use. And technologists created new machines in this positive meaning.

By contrast, in the twenty-first century, we must think in terms of Technica Negativa. Everything must be thrown away, thrown on the ground, after use: unfortunately, these things will often not degrade or return to the natural system. Waste will remain in the ground: this waste never fades nor becomes unified with the eco-system. If machine components could be returned to pure materials, we could use these materials again to realize new design by creating new machines. And we know that natural resources are very limited. For example, mineral resources on the earth are limited. We must think about an opposite or reversible process in the domain of creative activities in the factory. We must not only aim at recycling. Natural recycling must be reconceived in some sense, and we must give back materials into eco-system of nature. Unfortunately, the recycling process, which destroys and transforms used objects into different materials, often needs more energy than productive, new processes in the factory.

Is it possible for new technology to create more effective design by not recycling? Many questions arise.

(1) Are Technica Positiva and Technica Negativa contradictory concepts?
(2) If Technica Negativa is possible, it must be realized, like Technica Positiva, in the technological conjuncture or cohesion. How should we do that?
(3) Is Technica Negativa only a theoretical concept and not a practical one?

I consider Technica Positiva and Technica Negativa to be contradictory concepts not only in a theoretical sense but also in practice. Technica Negativa must be realized in the technological conjuncture. Therefore, we must think about the meaning of "Negation" or "Nothingness".

It is not a question of an antagonistic negation of technica and technology or technological sciences. In the medieval ages, we find the concept "theologia negativa", which constructs proof of divine existence—i.e., theologia negativa—as a theology of what God is not and becomes, so to speak, the integrated meaning of theology. On the other hand, Technica Negativa could construct an integrated meaning of technology. Perhaps, we should speak about the grounding of this negativity in a more profound nothingness. (cf. Imamichi 1968)

10.5 Nothingness

According to *Keys to the Japanese Heart and Soul* (Kōdansya Bilingual Books 1996), Nothingness is explained as follows:

> Fundamental Chinese Taoist concept, the term for absolute nonexistence, from which all existence arises and to which mankind ultimately returns. The idea of nothingness transcending ethics and politics was developed by Taoists critical of the Confucian

preoccupation with ethical and political problems. The philosophy of nothingness emerged with the Indian Buddhist philosophy of emptiness (ku) in Chinese and Japanese Buddhism, especially Zen Buddhism. As defined by Buddhism, however, nothingness is seen not as a state of nonexistence as opposed to existence but as an absolute, transcending the opposition of existence and nonexistence, or as an ideal and absolute human state identical to religious enlightenment.

In Oriental thought, Nothingness is one of the richest and the most important concepts for transcending the opposition between two antagonistic concepts; and, according to Taoists especially Zhuangzi, Nothingness is the origin from which all existence comes and to which all existence returns. It is an absolute. This means that negation is the integrated logical method to conceive a transcending cosmology.

I dare say that what is happening at the Fukushima Nuclear Facilities represents a concrete case of Technica Negativa. After the earthquake and following tsunami, the Japanese government was obliged to shut down all the nuclear reactors at Fukushima from No. 1 to No. 4 and begin demolishing them. But we do not have correct information about the state of the nuclear cores. It is certain that every core has melted down, but in what form? And we do not know what the structural situation of each facility is or the status of each nuclear core. Everything is completely destroyed. The inside of the nuclear facilities is full of radioactivity. The Japanese government has called for a new technology and new projects to research newly-developed technology for the world. It seems clear to the technologists that the most effective measure is "using robots". It is dangerous to come near the nuclear facilities. But, to this point, the Japanese government has received too few ideas, and the effort has been in vain. It makes no sense or it is not within our technological abilities. Around Fukushima, radioactivity is still measured at a fatal dose. But we must close these facilities. In the case of Fukushima, demolition is a very concrete example of doing Technica Negativa. The report of *Le Monde* on 18 December 2013 shows us that Fukushima is in the greatest demolition crisis we have ever experienced. It will take at least 40 years to demolish it. This tragic human accident will make for new developments in technology; but, as Kant suggested in the seventh thesis of his *Idea for a Universal History in a Cosmopolitan Sense*, because humanity is forced, due to suffering and evil, to institute a cosmopolitan condition to secure safety (cf. Kemp and Hashimoto 2011 and 2012).

References

Imamichi T (1968) Betrachtungen über das Eine, Institut der Aesthetik. Tokyo Universitat, p 228S

Imamichi T (ed) Revue Internationale de Philosophie Moderne, vols1–24, (1982–2009). Acta Institutionis Philosophiae et Aestheticae. Centre International pour l'Etude Comparée de Philosophie et d'Esthétique, Tokyo

Imamichi T (ed) (2003) Introduction to eco-ethica, special issue for the XXIst world congress of philosophy. Centre International pour l'Etude Comparée de Philosophie et d'Esthétique, Tokyo, p 102

Imamichi T (2009) An Introduction to Eco-ethica, translated by Judy Wakabayashi, University Press of America, Inc, p 102

Kemp P, Hashimoto N (eds) (2011 and 2012) Eco-ethica, vols. 1–2, Tomonobu Imamichi Institute for Eco-ethica, Copenhagen, Tokyo

Kemp P, Hashimoto N (eds) (2013) Eco-ethica, special issue for The XXIII world congress of philosophy in Athens, 2013 August. Introduction to eco-ethica, vol 3. Tomonobu Imamichi Institute for Eco-ethica, Copenhagen, Tokyo, p 99

Kodansya Bilingual Books (1996) Keys to the Japanese Heart and Soul. Kōdansya Bilingual Books, Tokyo, p 317

[illegible] (20[illegible]) [illegible]
[illegible]
[illegible]
[illegible]
[illegible]
[illegible]

Part IV
Models of Pre-design, Design, and Post-design

Chapter 11
Social Interactions in Post-design Phases in Product Development and Consumption: Computational Social Science Modeling

Russell Thomas and John Gero

Abstract This chapter presents a system to model social interactions between producers and consumers in the post-design phase. Producers form their expectations about consumer behavior during the pre-design and design phases. Consumers' behaviors are a result of their interactions with designs based on their experiences that form their value systems as well as their social interactions with other consumers. Because the post-design phase includes consumer behavior, producers reevaluate their plans and strategies for future designs. A subset of the system is implemented to model social interactions where the producers and consumers are modeled as computational agents. The agents' values that are used to guide their decision-making are modified through the agents' interactions with products and other agents. One of the goals of this work is to demonstrate the viability of agent-based modeling to study innovation ecosystems and their social aspects. Through computational experiments, we are able to test hypotheses regarding the mutual influence of producer and consumer values on the trajectory of design improvements. Exemplary results are presented.

Keywords Producers • Consumers • Social interaction • Agent-based modeling • Situated cognition • Producer–consumer interaction

---

R. Thomas (✉)
Krasnow Institute for Advanced Study, George Mason University,
4400 University Drive, MS 6B2, Fairfax, VA 22030, USA
e-mail: russell.thomas@meritology.com

J. Gero
Krasnow Institute for Advanced Study, George Mason University,
4400 University Drive, MS 6B2, Fairfax, VA 22030, USA

University of North Carolina at Charlotte, 9201 University City Blvd,
Charlotte, NC 28223, USA
e-mail: john@johngero.com

© Springer Japan 2015

T. Taura (ed.), *Principia Designae – Pre-Design, Design, and Post-Design*,
DOI 10.1007/978-4-431-54403-6_11

11.1 Introduction

In the eyes of a producer or designer, consumers often react to a new product introduction in surprising and unexpected ways. These unexpected reactions in the post-design phase are both a source of risk—products may fail in the marketplace—and a source of innovation as new uses or new utility are discovered, which in turn create opportunities for new directions in product design. Both producers and consumers form their expectations during the pre-design and design phases, including their experience with previous designs. But their behaviors and interactions in the post-design phase are not simply a matter of having those expectations met or not. In other words, design does not sit only between product requirements and product generation in the pre-design and design phases. Design also spans the post-design phase where there is often a complex interplay between *cognition*, *value systems*, and *social interactions* that reshape the design landscape and, cause producers to reevaluate their plans and strategies for future designs. This can be illustrated as follows.

Over the course of several product design generations, a producer will probably develop design competency in some areas (e.g. high volume production, narrow tolerances, ergonomic design) and not in others (e.g. reliability in extreme environments, high performance, small runs of custom configurations). Reinforced by success in the market, this producer will tend to value future designs that utilize the producer's competencies and exclude the design characteristics where the producer has little or no competency. Essentially, these design preferences are the producer's value system, which is also reflected in the producer's preferred stream of new product designs. We can view a stream of new product designs as a trajectory within the space of possible designs or within the space of possible performance/cost ratios. From their viewpoint, the Producer prefers new product designs that are similar to those where the Producer has the most experience and expertise, where it has relatively good profitability, and also where the Producer expects Consumer demand to be high or at least adequate. If consumers behave as the producer expects based on pre-design and design work, then the producer's strategic choice becomes one of choosing the optimal design trajectory and then executing it effectively.

But if consumers do not always react as expected or if they engage creatively with new and existing products in the post-design phase, then producers must reevaluate their strategic choices. This might mean abandoning existing competencies and preferred design strategies in favor of new and untried paths. (This is a form of Schumpeter's "creative destruction" Schumpeter 1942.) It might also mean that the producer might benefit from serendipitous events—e.g. product characteristics that were previously not valued by consumers suddenly come into favor, allowing a marginal producer to rise to market leadership. Producers that successfully observe and learn from consumers in the post-design phase can then adapt their strategies and plans in the subsequent pre-design and design phases, including the possibility of making fundamental changes in strategy or architecture.

Therefore the behavioral focus of our research is on how producers choose from alternative new product designs where the alternatives might differ in performance along dimensions related to consumer utility. We are especially interested in settings where the performance dimensions emerge endogenously rather than being fixed or introduced exogenously. For example, this would arise when new uses are found for a product, a use case is dramatically altered, or when new classes of users enter the market place.

To study this class of behavior and phenomena, it is necessary to observe the value systems and behaviors of both producers and consumers during the post-design phase, and preferably over several generations of product design to record changes in design trajectories. It is also useful to explore various experimental treatments to understand how alternative conditions of cognition, value systems, and social interactions affect outcomes. The phenomena of interest arise through the dynamic social interactions between agents, and between agents and artifacts, far from equilibrium. Therefore, it is important that the research method allows for the study of the endogenous processes for learning (direct and indirect), social interactions and network formation, and be capable of rich emergent phenomena. For these reasons, we have chosen to build a computational laboratory with multiple agent types, artifacts that agents produce and consume, and social interactions between agents. One of the main benefits of this research method is that many different experimental treatments can be explored and the detailed internal state of all agents is fully available for examination.

There is also a broader spectrum of interactions and forces at work in the post-design phase, including influences of government policies (e.g. intellectual property institutions), social interactions between producers (e.g. communities of practice), the value-shaping influence of third-part agents such as market analysts and 'gate-keepers', and the upstream influence of funding organizations and research institutions. Though these are beyond the scope of our current research, the architecture of our computational laboratory is extensible to include these other types of agents, artifacts, and interaction types.

One of the goals of this paper is to demonstrate the viability of Agent-based Modeling (ABM) to study innovation ecosystems and their social aspects. We have implemented a multi-agent system (Ferber 1999; Weiss 2000; Wooldridge 2008). It is designed to be a computational laboratory (Casti 1999) to support a wide variety of experimental settings and tests. Using multi-agent systems to simulate social systems is part of an emerging interdisciplinary field called Computational Social Science (Epstein 2012; Epstein and Axtell 1996; Gilbert and Conte 1995; Miller and Page 2009). Briefly, agents and their micro-level behaviors are formalized using relatively simple rules and limited/plausible capabilities for reasoning and behavior. An open environment is provided for agent interaction, often in the form of a grid or network. Through interaction with other agents and the environment, each agent alters its internal state, learns and adapts. The general research strategy is to study emergent phenomena that are not simple aggregations of the micro-behaviors (Gilbert 2002; Goldstein 1999; Holland 2000).

11.2 Theoretical Context

11.2.1 Co-evolution of Technologies and Consumer Preferences

Dosi (1982) introduced the idea of viewing technology evolution as trajectories through the space of possible designs, and movement along a trajectory as the result of "normal problem-solving" and "progressive refinement" by producers as they find ways to improve trade-offs in design variables. Saviotti (1996) presents a more formal model of technological evolution through design space, where the space is defined by dimensions for each technical and service characteristic associated with a particular technology. 'Characteristics' are formalized as a vector of variables that specify both a product's internal structure ('technical characteristics') or services performed for its users ('service characteristics') (Saviotti and Metcalfe 1984). This "twin characteristics framework" is important for understanding both the producer's values, which center on technical characteristics and associated learning, and the consumer's values, which center on the service characteristics. We apply this method for modeling the space of possible designs and to specify the position of particular product designs within that space. Gero (1990) is a further elaboration of these ideas in the field of design science, specifically via Gero's Function-Behavior-Structure (FBS) ontology for designs. 'Structure' in FBS corresponds to Saviotti's 'technical characteristics', while 'Function' corresponds to Saviotti's 'service characteristics'. 'Behavior' provides the ontological linkage between Structure and Function, and thus are often the focus of attention of product designers in pre-design and design phases. Because our simulation does not include agents actually performing design acts or making explicit design decision, we do not explicitly include Behavior in our model of product designs.

Saviotti (1996) also proposes methods of analyzing population-level dynamics in design space such as movement along trajectories and changes to the 'technological frontier'. The latter is related to the 'adjacent possible', a phrase coined by Kauffman (1996) for the set of all the designs that are directly achievable from an existing set of competences. Thus, the technological frontier is the limit of what is producible with today's costs and capabilities, while the designs in the set of the 'adjacent possible' are decision alternatives for producers who choose to expand or extend the frontier.

Dosi and Nelson (2010) provide a recent survey of the state of research on technology trajectories and evolutionary processes that give rise to them. They describe the supporting evidence as "ubiquitous", adding that "trajectory-like patterns of technological advance have been generally found so far whenever the analyst bothered to plot over time the fundamental techno-economic features of discrete artifacts or processes." (p. 16) Technology trajectories are an example of emergent phenomena (Gilbert 2002; Goldstein 1999; Holland 2000) in that they arise from the collective action of individual agents but are not simple aggregations of agent behaviors. As Dosi and Nelson (2010) describe, technology trajectories have a downward causal influence on agents, effectively circumscribing technological advances "within a quite limited subset of the techno-economic characteristics space. We could say that the paradigmatic, cumulative nature of technological knowledge provides innovation avenues (Sahal 1985) which channel technological evolution, while major discontinuities tend to be associated with changes in paradigms."

Compared to the large literature on technological evolution and trajectories, there has been much less research on the co-evolution of technologies and consumer preferences/values. Evidence of this is given the survey of the state-of-the-art in Dosi and Nelson (2010). They discuss research on how demand and other socio-economic factors shape the direction of technological advance, but they do not cite any research that specifically focuses on endogenous co-evolution of technology and consumer preferences. However, there has been research on specific topics related to co-evolution of technology and consumer preferences/values, for example: the role of experimental users and diverse consume preferences (Malerba et al. 2007), consumer resistance to innovations (Moldovan and Goldenberg 2004), compatibility and innovation (Sosa and Gero 2007), and innovation as changes in value systems (Gero and Kannengiesser 2009).

Separate from technological evolution, there is an extensive literature on consumer preferences, opinions, and consumer behavior. Liggett (2010) evaluates alternative methods for mapping consumer preferences as a population using perceived product characteristics and their 'ideal product' which can be formalized as a vector of values for each service characteristic of the product. Liggatt also uses Multi-dimensional Scaling (MDS) to create a 2D map of a population of consumers' ideal vectors relative to the available products. There is also extensive research on how consumers influence each other's values and opinions through social interactions, e.g. Friedkin and Johnsen (1999). This literature guided our design decisions for social influence mechanisms and patterns, including topology of consumer social networks, the behavior of opinion leadership, susceptibility to social influence, and homophily as a primary factor in the determining strength of social ties and thus the degree of influence between any two consumers.

11.2.2 Situated Cognition and Innovation

Situated cognition (Clancey 1997) provides the theoretical basis for our design of agent cognition. Any cognitive system operates within its own worldview and that worldview affects its understanding of its interactions with its environment (Clancey 1997; Gero 2008). In essence, what you think the world is about affects what it is about for you.

A person or group of people is 'situated' because they have a worldview that is based on their experience (Smith and Gero 2001). Situated cognition involves three ideas: situations, constructive memory and interaction. Situations are mental constructs that structure and hence give meaning to what is observed and perceived based on a worldview. Constructive memory makes memory a function of the situation and the past. Memory is not laid down and fixed at the time of the original sensate experience. What is remembered is constantly being recreated and reframed. Interactions between agents trigger changes in situations and memories.

Through the lens of situated cognition, innovation in a social ecosystem is an emergent phenomenon that arises from the interplay of situations, constructive memory, and social interactions at the level of agents and networks of agents. Moreover, we believe that situated cognition is at the heart of social processes of creativity and inventiveness (Gero 1996; Sawyer and Sawyer 2012). This is especially relevant to emergent aspects of innovation (Finke 1996; Gero and Damski 1997; Gero 1996; Gero and Kannengiesser 2004; Sawyer and Sawyer 2012). For these reasons, situated cognition is foundational for any study of innovation in social ecosystems (Edquist 2005; Llerena 2006).

In the current implementation of our simulation, only Consumers are designed with features to support situated cognition. In future work, we expect to add situated cognition capabilities to Producers and other agent types.

11.3 Architecture of the Simulation System

There are two types of agents in the current implementation—'Consumers' and 'Producers'—and one type of artifact—'Products'. Throughout each simulation run the population size is fixed for Consumers, Producers, and Products. The population size is under experimental control and can range from ten to hundreds or thousands, limited only by computer capacity and processing speed. To date, our experiments have been run on a MacBook Pro laptop with 2.53 GHz dual core processor and 4 GB main memory, with a population of 100–200 Consumers and a similar number of Products.

Consumers seek to consume Products by moving around a geographic Consumption Space with micro-behavior similar to foraging, but with social interactions. Consumers are not endowed with any knowledge or map of the Consumption Space, nor do they have any memory of where they have been. The Consumption Space is a bounded rectangular grid with von Neumann neighborhoods, and the size of the Consumption Space is proportional to the population size of Products and Consumers so that the spatial density remains constant. This facilitates foraging micro-behaviors and minimizes behavioral artifacts that might result from a physical landscape that was either too dense or too sparse. In each clock cycle consumers can move to any neighboring point on the grid within the boundaries. At each time step Consumers move around the landscape looking for attractive Products to consume, or to maneuver out of crowded areas. Only one Consumer can occupy a given grid location at a given time.

Consumers are social, while Producers are not. The social network among Consumers is initialized as a 'small world' network with random assignments.

Once the simulation starts, Consumers form new social relations when they meet each other or by referral through their existing social network. The initial strength of a social tie is proportional to the similarity between the two Consumers. Strengths of social ties decay with time unless they are recharged by exchanges of information. While the structure of the social network evolves endogenously, when ever a Consumer has no social ties the simulation system intervenes and creates several new social ties.

Products are initially distributed randomly in the Consumption Space, where they remain until they are consumed or they expire. If they are consumed or expire, they are replaced by the Producer(s) but not necessarily with a Product of the same type. During simulation initialization, a relatively large number Product types are generated by the system and these comprise the 'Product Design Set', i.e. the set of possible new Products that may be introduced during the course of a simulation run. A small subset of these possible Products are selected as the initial set of 'active' Products. Each new Product type is selected from the 'adjacent possible' relative to the currently 'active' Product set. Each new product introduced expands the 'adjacent possible' to include Products that were previously not feasible for Producers. If the Product Design Set is large relative to the length of a simulation run and the rate of new Product introductions, then we are able to simulate a continuous stream of innovations.

Though Producers have several cost measures associated with each type of Product, there is no price to Consumers and thus no financial transaction between Consumers and Producers. This is because our research focus is on the interaction of Consumer and Producer value systems and not on profit-maximizing decisions given limited resources.

Producers only take action when a Product is consumed or expires. When that happens they make a decision to either replace it with an identical Product type, a different Product type in the portfolio of available types, or to introduce a new Product type that had previously not been available. Producers have no direct interaction with or knowledge of individual Consumers, therefore they make their decisions based on historical data regarding the consumption and expiration Products of various types. Also, in the current implementation, Producers have no social interactions with each other, nor any knowledge that there are other Producers. In the results presented in this paper, the simulation was configured to have a single Producer.

The population of Consumers initially have identical values and other initial conditions, and are spaced randomly with uniform distribution on the Consumption Space. The adaptive elements in their cognitive architecture are primed in two ways. During initialization Consumers repeatedly sense, perceive, and evaluate all of the products in the initial 'active' set. Then the simulation is started in a 'priming mode', which is fully functional except that experimental data is not collected. Priming is complete when all or nearly all Consumers are sufficiently adapted to make consistent valuation and consumption decisions. After priming is complete, the Consumption Space is reset to it's initial conditions for Products and Consumers, including the social network.

11.4 Computational Modeling of Agents and Artifacts

11.4.1 Products

Products are constructed as a graph structure with six nodes. During the construction process, edges between the nodes are formed at random, creating a single connected graph with between 5 and 14 edges. There are 112 unique six node connected graphs, yielding a design space for agent exploration that is small enough to be tractable for enumeration and complete analysis.

Products have both surface characteristics and functional characteristics. During their search and evaluation process, Consumers can only sense and perceive a Product's surface characteristics (its 'signature'). The functional characteristics are only experienced through the process of consumption. During consumption, Consumers gain utility based on the functional characteristics, relative to expectations. Higher than expected evaluations of functional characteristics yield positive utility, while lower than expected evaluations of functional characteristics yield negative utility. The surface characteristics of Products are related to their functional characteristics, but not identical. Consumers cannot directly perceive the utility of Products.

The Product's utility is a function of its topology, while its external appearance is a function of its physical layout. Physical layout (Fig. 11.1) is constructed on a circle with a relaxation method to equalize the length of edges. Distance from the centroid of this layout produces a signature in the form of a six-element vector. Four more elements are added based on other surface characteristics of the graph, for a total of ten elements in a signature vector.

Three dimensions of Consumer utility have been defined (Fig. 11.2) in a way that might be analogous to performance dimensions in a manufactured component. The first is the number of edges, which ranges between 5 and 15, and might be analogous to a property like weight or density. The second is mean clustering coefficient, which might be analogous to a property like stiffness/flexibility. The third dimension is the maximum distance between nodes in a spring-electrical embedding. This might be analogous to maximum span distance.

The construction process for Products yields a non-obvious relationship between the surface characteristics and the functional characteristics of Products. Products that are close in surface characteristics (i.e. close in Value Space) might have different utilities. This allows for a somewhat rugged landscapes.

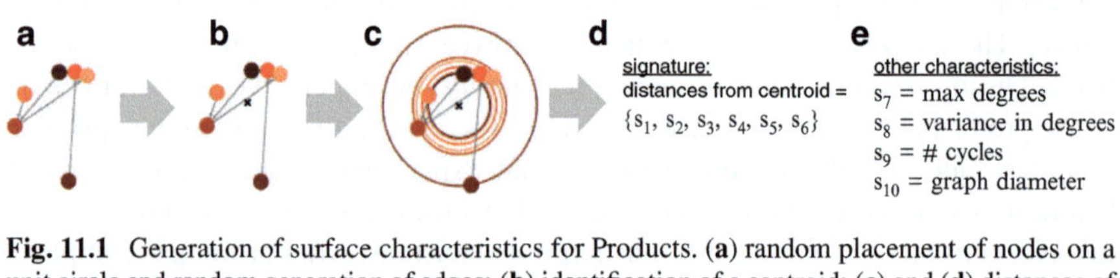

Fig. 11.1 Generation of surface characteristics for Products. (**a**) random placement of nodes on a unit circle and random generation of edges; (**b**) identification of a centroid; (**c**) and (**d**) distances of nodes from centroid determines the signature; (**e**) other physical characteristics available to Consumers through sensation and perception

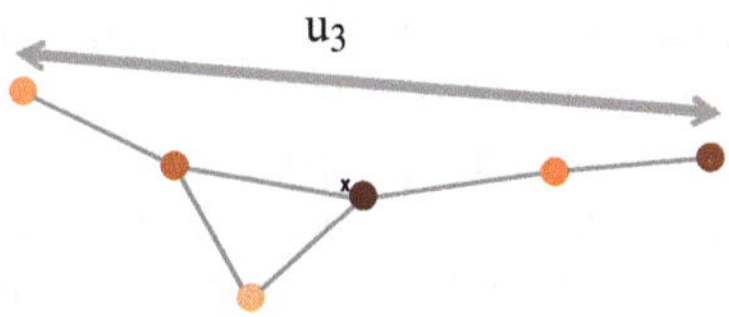

u_1 = # edges
u_2 = Mean clustering coefficient
u_3 = max distance
in spring-electrical embedding

Fig. 11.2 Maximum spanning distance is the third of the three dimensions of utility, measured on the same Product as shown in Fig. 11.1. The normalized utility vector of this product is (0.1, 0.39, 0.46)

In addition to these two views that are relevant to Consumers, we also characterize Products in ways that are particularly relevant to Producers. This is an important feature to our design because of the need to model plural interests, perceptions, and value systems between Producers and Consumers. We have adapted the idea of 'production recipe' from (Auerswald et al. 2000), where it was used in an agent-based model of learning-by-doing on the shop floor. A production recipe is a vector of characteristics that related to the production or assembly process, and therefore to the costs and complexity of manufacturing and the challenges of learning through experience. We have defined the following eight-element vector for specifying recipes:

r_1 = Number of degree-1 nodes
r_2 = Number of degree-2 nodes
r_3 = Number of degree-3 nodes
r_4 = Number of degree-4 nodes
r_5 = Number of degree-5 nodes
r_6 = Log of number of cycles = $ln(c)$, rounded to nearest 0.5
r_7 = Length of longest chain
r_8 = Number of chains of degree-1 or degree-2 nodes

These are each normalized to a range of values between 0 and 6. There are 91 unique recipes for the 112 unique Products. The Hamming distance between any two recipes is a measure of accessibility from one to the other through learning-by-doing and also explicit design explorations.

The cost to manufacture a given design has two components. The first is material, which is a simple function of the number of edges. The second is assembly cost, which is a function of the recipe and the Producer's cumulative experience in each of the dimensions of the recipe. With zero experience, the cost function rises as the square of each recipe element value, summed across the recipe. Thus, initially most of the designs are too expensive to manufacture, rendering them infeasible. With experience the exponent of the cost function is reduced until it plateaus to yield a linear function of each recipe element value.

An important feature of the production recipe approach is that Producers can gain experience in one set of Products that lowers its costs for other Products in the 'adjacent possible' region of design space. However, because the Producer does not have full knowledge of the design space, the trajectory of design choices emerges through a series of local/limited decisions, adaptations, and also constraints of attention. It is not governed by foresight or planning.

11.4.2 Consumers

Consumers search the landscape for attractive Products to consume, and they form social networks in the process. Consumers modify their values through direct product interaction (evaluating and consuming Products) and through social interactions. Consumption decisions and subsequent learning are mediated by two independent variables—'value' and 'utility'. 'Value' is the Consumer's appraisal of a Product based on its surface characteristics, relative to that Consumer's ideal. Thus, valuation is performed prior to any consumption decision. Consumers choose to consume based on their perception of product signature, perception of proximity to their ideal type, and a rough expectation of utility. Generally, Consumers choose to consume when the Product they encounter is close to their ideal type. The space of possible Product signatures is called the Value Space. The value system for each Consumer centers on a single vector that represents the signature of its ideal product type. Consumers learn and adapt by adjusting this vector through experience and social interaction. Therefore, each Consumer's value vector can be represented as a point in Value Space, and their changing values as paths through Value Space.

In contrast, 'utility' is the benefit that the Consumer receives after consuming the Product. It can only influence future consumption decisions through agent learning and, indirectly, through social interactions. In the current implementation, there are three possible utility functions (Fig. 11.2).

Figure 11.3 shows a simplified block diagram of the agent architecture. Compared to other agents in the social network influence literature, these agents have a rich architecture that includes both symbolic and sub-symbolic reasoning. This was necessary to implement situated cognition, which was one the primary research goals. Due to space limitations, we will only describe perception, conception, situation, valuation, and social interaction functions.

Perception is the collection of functions that enable the agent to focus and organize their sensations according to their current situation, their expectations, and past experiences. Consumers perceive Product signatures using a Self-organizing Map (SOM, also known as Kohonen Networks). SOMs are a type of neural network that are trained via unsupervised learning. Essentially they perform a mapping from the sensed Product Signature to a condensed 2D internal representation of the Products. This is functionally equivalent to conceptual spaces, as described in Gärdenfors (2004). Perception is updated every step, but is only processed when new sensations arrive.

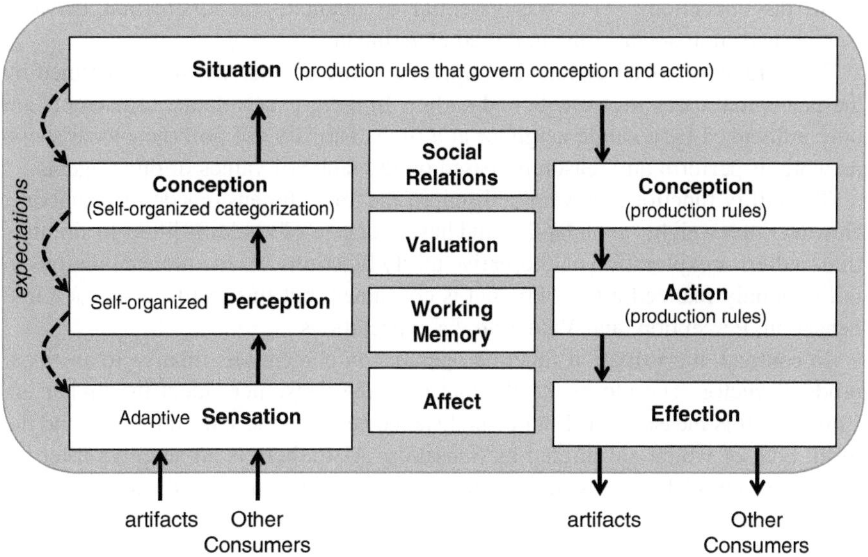

Fig. 11.3 Simplified agent architecture for Consumers

Perception of other agents is performed through a categorization and comparison procedure, where agents with direct social connections are labeled 'most similar', 'most dissimilar', 'most admired', 'least admired', etc.

Conception is the highest level of reasoning, including both tacit and explicit capabilities. Tacit conceptual reasoning is focused on deciding whether a given product should be considered attractive or not. This is implemented using a SOM that essentially creates a one-dimensional map of products that it has experienced and the value and/or utility that is perceived or realized from those products. A threshold value is used to trigger a decision that the product being inspected is attractive. The threshold value is adapted through learning by experience. Conception also includes explicit reasoning about actions, social interactions and goals. These are implemented symbolically as 'If-Then' production rules.

Situation has the function of cognitive orientation and focus. We implemented it using production rules that test for conditions that would change a Consumer's situation, and then fire actions to change their concepts accordingly. Overlapping situations can be active at the same time. Situations act as conditions on other conception rules so that each conception rule fires only when one of its applicable situations is active. Situation also activates other reasoning functions, as appropriate.

Social interactions are implemented using production rules. Generally, social interaction only occurs when the Consumer is both not in the act of consumption and is also frustrated by its consumption experiences. The exception to this occurs when a Consumer's social ties have been reduced to two or fewer and their strength has fallen below a threshold value. Here, conception rules fire that cause the agent to create new social ties. This is necessary in order to sustain social networks and avoid disconnections. This was essential to maintain the distinction between 'social' and 'non-social' runs in paired experiments.

The targets of social actions and influences interactions are always defined by the perceptual categories mentioned earlier. In these experiments, Consumers are only influenced by a single neighbor at a time. They do not poll their local social network or perform any reasoning based on the range of values of other agents.

The utility function is a weighted sum of the three dimensions described above. However, the weights are adaptive and have a degree of random 'jitter' to simulate trial-and-error exploration of alternative utility functions. As mentioned previously, utility is only realized after a Product is consumed, but this result feeds back into Sensation, Perception, and Valuation via expectations.

In contrast, the valuation function operates on perceptions relative to an 'ideal product' vector. The closer the Product is to the consumers ideal the higher the valuation. It is measured as Euclidean distance between the sensed product and the ideal, both of which are filtered by Sensation. Also, there is an adaptive filter for valuation to model agent focus and prioritization, and also generalization.

In summary, at the task level the Consumer's problem is to find relatively more desirable Products to consume by searching the Consumption Space and adjusting their ideal product type. If they become dissatisfied during this process or if they are not able to find products to consume, they interact socially to either modify their value system or to move toward another agent in the Consumption Space.

At a social level, Consumers create and maintain social relationships through physical contact in the Consumption Space. However, if a Consumer is close to losing social connections, that Consumer interacts socially to build new connections through a referral process ('friends of friends'). The focus of social interaction is on soliciting or offering information about another Consumers' ideal product. We simulate the phenomena of opinion leadership and also susceptibility to influence from others.

11.4.3 Producers

In the current implementation, Producer agents have a simple architecture focused on two decision processes. They use simple decision rules based on local optimization and, unlike Consumers, do not have any other cognitive capabilities for sensation, perception, conception, or affect. The only decisions they make are (1) current production—choosing a replacement product from existing designs to replenish inventory in response to Consumer acts of consumption, and (2) new product introduction—choosing a new product designs to introduce from the designs that are in the 'adjacent possible'. To decide on current production, the Producer uses consumption statistics by region, and then makes their choice from existing designs based on weighted random choice. Weights are set to be proportional to consumption history in that region. Also, a fixed weight is given to recently introduced products to encourage their selection and production, even though they may not have much history of consumption in that region.

New product designs are initially given a ranking according to their performance/cost ratio rank in three dimensions of performance. We assume that product designs are introduced in a sequence of improving performance/cost ratios, starting from the lowest and culminating in the highest, yielding a trajectory of product designs in the space of performance and cost. Since there are three performance dimensions, there are multiple possible trajectories for any given set of possible designs, and the trajectories may or may not be disjoint.

Because of feasibility and cost constraints, Producers will initially produce the designs with lowest costs (both design and manufacturing) and therefore relatively low (unattractive) performance/cost ratios. Subsequent new product introduction choices are made from the designs that are next in sequence of performance/cost on any of the performance dimensions. This defines the 'adjacent possible' for the Producer. When more than one product design is in the 'adjacent possible', Producers face a strategic decision to either stay on their current design trajectory or to move on to a new trajectory. Where design trajectories diverge ('branching points'), Producer choices for which new product to introduce determine which design trajectory is realized and which are not. This creates path dependence in Producer values and, indirectly, in Consumer values too.

11.5 Method

11.5.1 Experimental Design

Our goal is to study how Producer's choices from alternative designs are affected by evolving Consumer demand and preference (in the post-design phase) as revealed by their consumption behavior. Producer's choices should reveal how their post-design learning influences their actions in succeeding pre-design phases.

During the course of a run, Producers can only directly affect their costs by learning-through-experience. To increase product performance (i.e. to offer higher utility to Consumers) Producers must discover and offer new designs as they become feasible and cost-effective. On the other hand, Consumers can only influence their utility through consumption decisions, which may be satisfying or not, and by modifying their value systems so that the Products that are available are better appreciated (possibly).

The phenomena of interest are design trajectories in the space of Product types, as measured by performance/cost ratios over time. Divergences between alternative design trajectories represent discontinuous change and (potentially) disruptive innovations. The simulation system has been designed to allow, from identical initial conditions, different design trajectories can be realized depending on the co-evolution of Producer and Consumer values.

Therefore, our experimental approach involves comparing results across three experimental settings with identical initial conditions for the Product Design Set:

1. A single Producer acting in isolation from Consumers, with a deterministic consumption rule.
2. Consumers acting in isolation from Producers, with a deterministic Product replacement rule.
3. A single Producer interacting with a population of Consumers, with endogenous consumption and production/innovation processes.

The first setting allows us to measure the outcomes of Producer's decision process considering only the needs and values of the Producer. We expect to see design trajectories that are chosen based on local optimization and accumulated 'learning by doing', given the initial 'active' Product set and the topology of the Product Design Set. We can consider these realized design trajectories to be 'preferred' by Producers in the absence of other influences. Furthermore, we measure the Producer's value system by the weights they assign to the three alternative utility measures.

The second setting allows us to measure the outcomes of Consumers' decision processes considering their needs and values alone. We expect to see Consumer values—i.e. their 'ideal product' vector and their utility weights—adapt continuously as new products are introduced. However, Consumer adaptation will stop when they find Products that are repeatedly and sufficiently satisfying, and thus 'make up their mind' by resisting any further changes to their values. This should lead to two types of results. First, the population of Consumers will become segmented as sub-groups form with similar values that are resistant to change. This will be visible as clusters in Value Space and Utility Space. Second, there will be considerable path dependence in trajectories in Value Space and Utility Space depending on the initial 'active' Product set and also the sequence of product introductions. This will mean that if Products are introduced in the 'wrong' sequence, Consumers may not find them attractive enough to consume, even though they would do so in other, more favorable circumstances (Saunders and Gero 2004).

The third setting allows us to measure the value system changes that result from the mutual influence of Consumers and Producers. If design trajectory outcomes in this third setting resemble those of the first setting, then we can say that the Producer's values have not been significantly affected by the Consumer's behaviors and their changes in values. Likewise, if the trajectories and patterns of Consumer values in Value Space and Utility Space resemble those from the second setting, then we can say that Consumer values have not been significantly affected by Producer decisions and changes in their values.

However, we expect to see significant differences in the third setting compared to both the first and the second, which would provide evidence that both the Producer's and Consumers' values have a strong mutual influence on each other that is not predicted by their behavior in isolation. Furthermore, this sets the stage for possible emergent phenomena and 'collective intelligence' that arise from this mutual influence.

11.5.2 Treatments

The primary experimental treatments are different rules for the initial set of Products available for consumption (the 'active set') and the agent's rules for making changes. For the Producer, the key decision is when to introduce a new Product from the available alternatives. This depends on their value system—how much they value 'profitability' (i.e. high performance/cost ratio), how they weigh the three dimensions of Consumer utility based on consumption history, etc. For Consumers, the key rule is how quickly or easily they 'make up their mind' about what product characteristics and utility dimensions they prefer, and how open they are to change and influence after that. While we will initially experiment with a few alternatives, eventually we intend to run parameter sweeps across a range of Product Design Sets and agent value system rules.

For each run of the experiment, we run each of the three experimental settings with the same initial conditions for the Product Design Set, which includes both an initial set of Products available for consumption (the 'active set') and the set of possible new Products covering improvements in performance/cost in one or more dimensions. Each treatment will be run repeatedly with other initial conditions randomized. The resulting design trajectories and value system trajectories are analyzed statistically to identify differences between the three settings.

11.6 Results

We are in the early stages of experimentation and therefore we present preliminary and illustrative results in this paper.

11.6.1 Setting 1: Producer Acting in Isolation

The following describes illustrative results in the first experimental setting—Producer acting alone.

The initial 'active set' of products are those that have an initial manufacturing cost below a fixed threshold, chosen so that it would exclude all but a few products. Figure 11.4a is a histogram of initial manufacturing costs and Fig. 11.4b is a histogram from a single run after 3,000 steps. Costs are high initially because the Producer has no experience and thus faces a cost function that rises with the square of recipe element values.

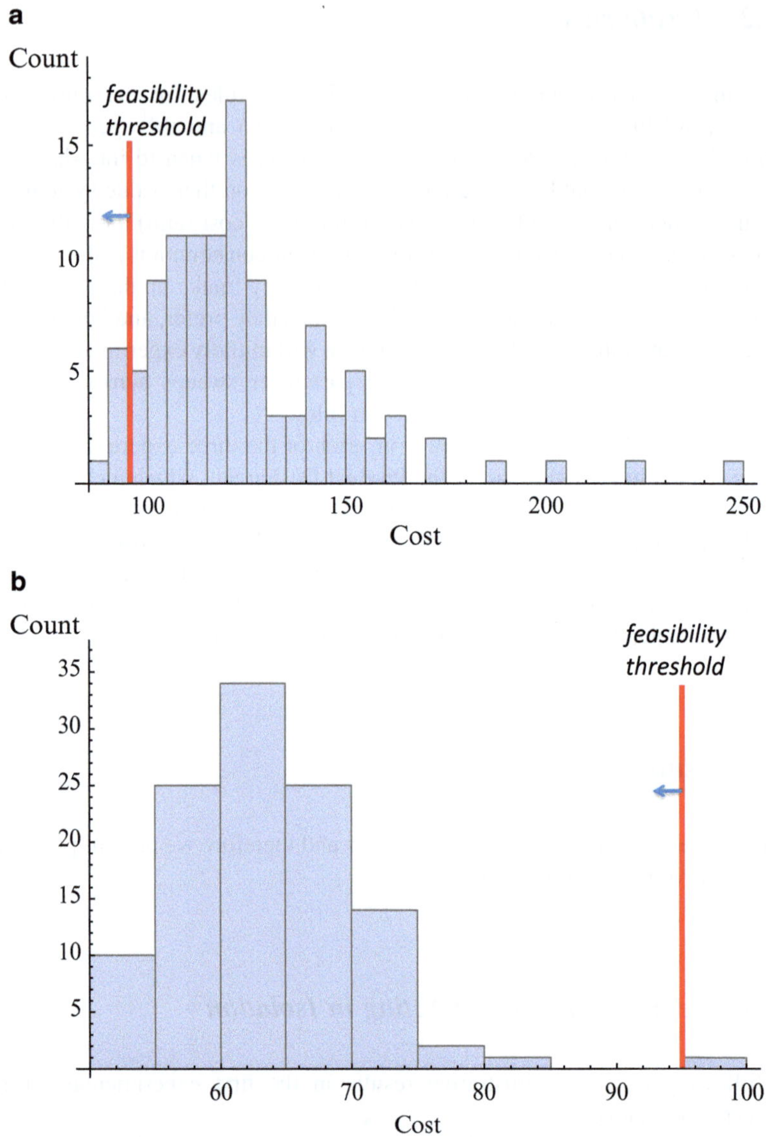

Fig. 11.4 Histogram of manufacturing costs for all possible designs with Producer using Rule 3—"Maximize performance/cost ratio" (**a**) $t = 0$ (**b**) $t = 3{,}000$

In this setting, the Producer manufactures a fixed quantity of product each step in the simulation, but that quantity is weighted across the 'active set' according to the Producer's value system. There are three weighting rules that we have explored:

1. *Minimize cost*—essentially focused on exploiting the benefits of learning-by-doing.
2. *Maximize performance*—a proxy for providing the most utility to Consumers without regard to cost as long as it is below the feasibility threshold.
3. *Maximize performance/cost ratio*—a middle ground between (1) and (2). This corresponds to profit maximization in micro-economic models.

The graphs in Fig. 11.5 show cumulative production by product design index for a single run for each of the three rules but with the same initial conditions. The stopping condition for each run was either 3,000 time steps or producing 500 units of the design with maximum utility under that rule. Thus the production quantities are different between runs because they reached the stopping condition at different times.

The diagrams in Fig. 11.6 show how design trajectories differ under the three Producer rules. All 112 possible designs are shown in the space as a grey dot (potential), or other dot as described in Fig. 11.6. The layout of the space is suggestive of proximity between designs from the Producer's point of view—i.e. proximity of production recipes. This layout was created using distance to 12 nearest neighbors and a graph layout algorithm—spring-electrical embedding. Other methods were tried, including Multi-dimensional Scaling, but none produced

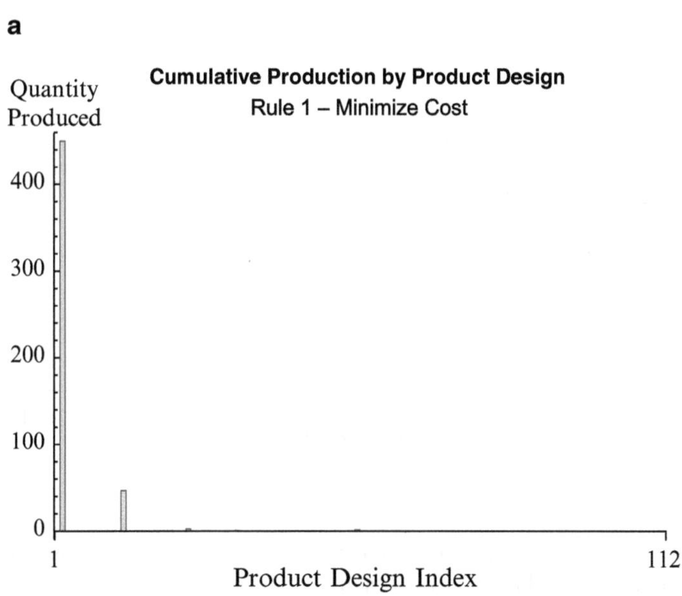

Fig. 11.5 Results of three runs with different valuation rules for Producer acting alone (Setting 1) (**a**) Rule 1 (**b**) Rule 2 (**c**) Rule 3

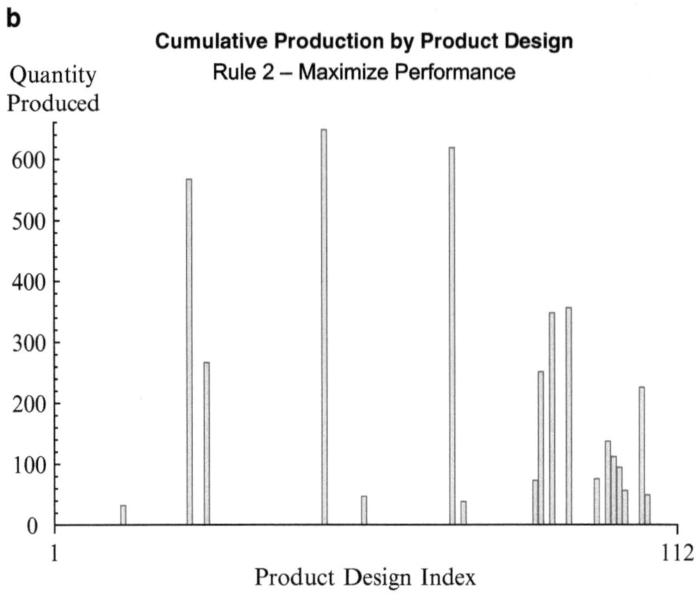

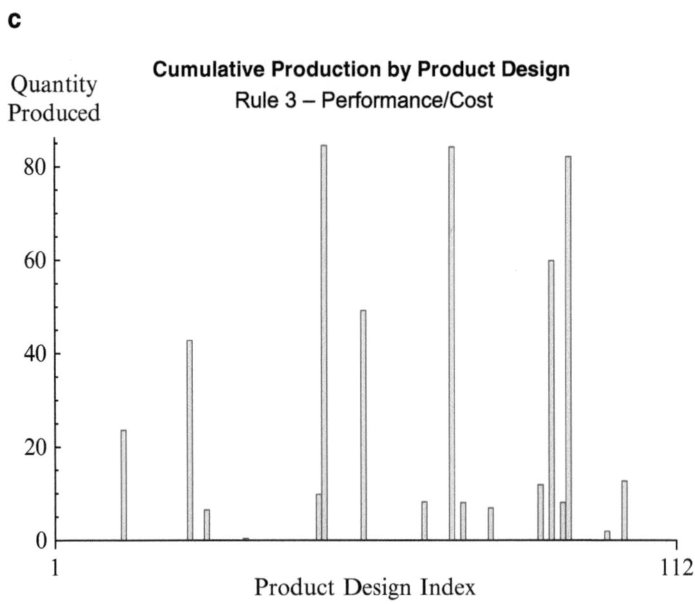

Fig. 11.5 (continued)

results that were as good as this for visual clarity. The light blue lines and dots show the relevant neighbor relations for the 'active set', i.e. the adjacent possible. Notice that many of these lines span a long distance indicating that the true dimensionality of the design space is high.

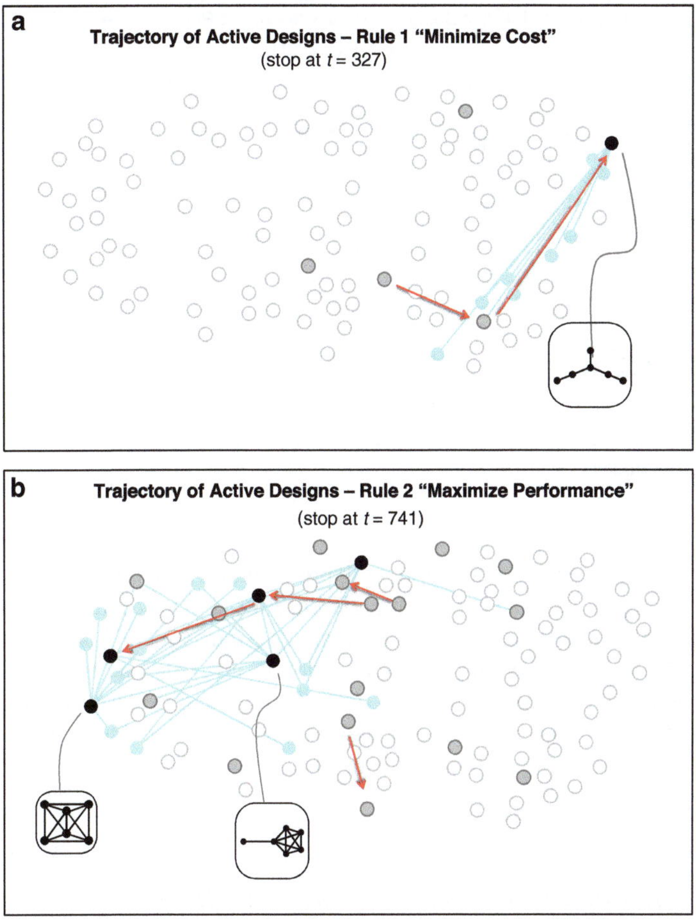

Fig. 11.6 Design trajectory from a single run for three different Producer rules. Key: *black dot* = active set; *grey dot* = previously active but no longer produced; *light blue dot* = 'adjacent possible' designs; *hollow grey dots* = all possible designs; *red arrows*: the main trajectory (**a**) Rule 1. *Inset*: Product design #2 (**b**) Rule 2. *Inset*: Product design #108 (*left*) and #100 (*right*) (**c**) Rule 3. *Inset*: Product design #105 (*left*) and #28 (*right*)

These diagrams reveal three broad differences in trajectories. First, the speed of movement along the trajectory is very different. Rule 1 moves very quickly to a fixed point where it only produces Product #2. We suspect that this pattern will be repeated with many other initial conditions, though maybe not all. Rule 2 moves fairly quickly toward the region of design space that is populated by the most complex designs. But Rule 3 meanders and moves more slowly, even oscillating between brand new designs and revivals of previous designs.

Second, the destination of the trajectories are clearly different. Rule 1 seems to gravitate toward simple designs, while Rule 2 does the opposite. As might be

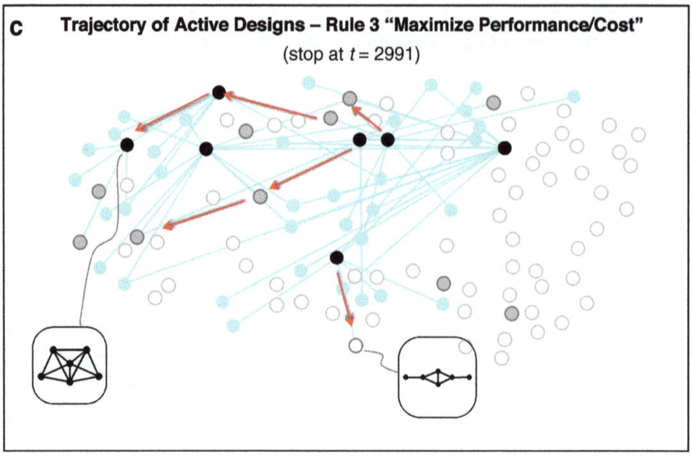

Fig. 11.6 (continued)

expected, Rule 3 seems like a hybrid, with predominant preference toward the high complexity region but also persistent preference for simpler designs.

Third, the number and diversity of product designs varies. Rule 3 is the most diverse, and seems to have a strong inclination to diversify and try many designs simultaneously, even previous designs. Rule 1 is the least diverse, and Rule 2 is in between. In our future experiments we will be measuring these differences in trajectories to test our hypotheses statistically.

11.6.2 Setting 2: Consumers Acting in Isolation

When testing Consumer behavior separate from Producer behavior and decision rules, our goal is to present them with relatively simple patterns of change and look for endogenous and emergent responses. Thus, we are presenting them with somewhat arbitrary design trajectories and we are looking for general patterns regarding how they respond to or resist the changes.

To analyze Consumer behavior and changes in their value system, we use two representations not present in the analysis of Producers. The first is 'Value Space'—a Multi-dimensional Scaling (MDS) map of Consumer ideal product vectors along with the characteristics vector of each Product (both active and potential) (Liggett 2010). In the graphs shown below, these are unfiltered by Sensation. The second representation is 'Utility Space'—a simple three-dimensional map of each Consumer's utility weights in a triangular barycentric coordinate system.

One interesting general pattern is clustering of values and preferences. We would be interested in knowing if clusters (a.k.a. 'market niches') form

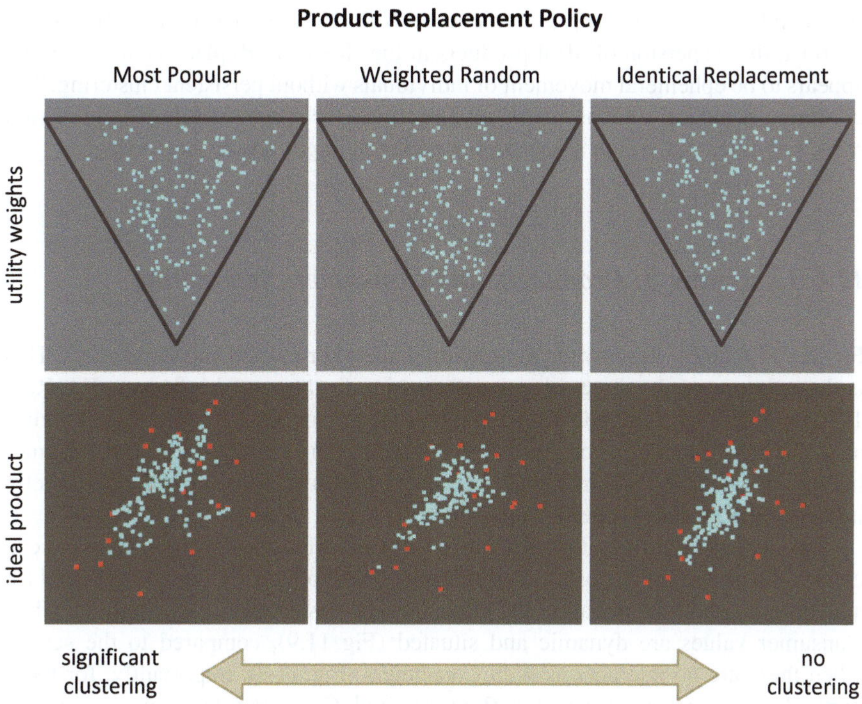

Fig. 11.7 Results from three runs with different treatments but otherwise identical initial conditions. Clustering of Consumer values is only sustainable if Producers are responsive to their initial preferences. Key: *Blue Dots* are individual consumer ideal vectors (*bottom*) or utility weights (*top*). *Red dots* are Products in the active set

endogenously and are sustained. To test for clustering, we implemented three simple production rules (really, product replacement):

1. *Identical Replacement*—the same Product replaces a Product that is consumed. Thus the mix of Products never changes.
2. *Weighted Random Replacement*—replacement products are chosen by random draw with the probability being proportional to past consumption in that broad region.
3. *Most Popular*—the replacement product is chosen to be what ever the most popular product (most consumed) in that region.

The images in Fig. 11.7 show the results of these three alternative treatments for a given random realization, for Steps = 6,000. The three treatments differ in how responsive the Producer is to Consumer demand patterns. 'Most Popular' is the most responsive policy, followed by 'Weighted Random', which is somewhat responsive, and followed by a non-responsive policy of 'Identical Replacement'. The most significant result is the degree of clustering in Value Space (bottom row) for the responsive policies. These clusters persist over extended time periods (~100 steps) and establish themselves in niches in Value Space far from the centroid. In contrast, the dispersion of ideal products in the 'Identical Replacement' treatment appears to be ephemeral movement of individuals without persistent clustering. The clustering behavior is not as visible or apparent in the space of Utility Weights (top row), but other analysis or visualization methods might reveal some effects.

11.6.3 Setting 3: Producers and Consumers Interacting

Figures 11.8 and 11.9 show the mutual influence of Producers and Consumers. This is the only one of the three rules that show significant mutual influence. For Rule 1—Minimize cost, adding Consumers interaction made no difference since it didn't alter the cost function. For Rule 2—Maximize performance, adding Consumer interaction also had no influence because Consumer demand did not directly influence the Producer's performance metric.

The situation is different for Rule 3 because variations in Consumer demand can change the performance/cost ratio as measured by Producers. Comparing Figs. 11.8 and 11.9, it can be seen that movement along the design trajectory is slower when Consumer values are dynamic and situated (Fig. 11.9), compared to the setting when there are no interactions with Consumers (Fig. 11.8). Importantly, for these particular parameter settings for Producer and Consumer learning, the design trajectory is not altered by Consumer interaction. However, that does not rule out the possibility that there are parameters values that can result in a dramatic shift in trajectories. Using full parameter sweeps, we will look for critical parameter values in future research.

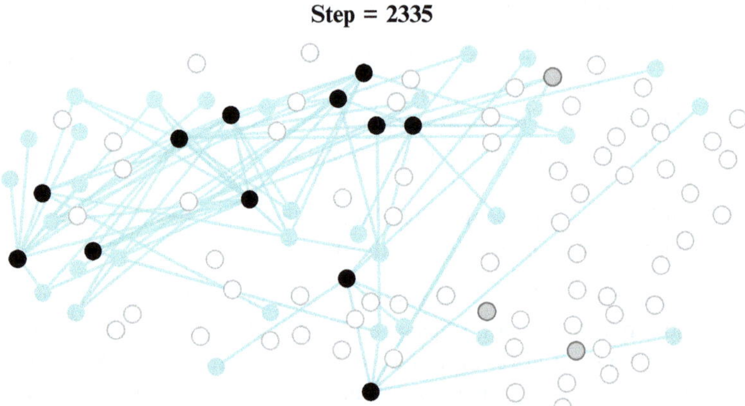

Fig. 11.8 A snapshot for Setting 1—Producer only, showing the active set of designs (*black dots*) using Rule 3—Maximize performance/cost ratio, recorded when the target metric reached 25

Fig. 11.9 A snapshot for Setting 3—Producer–Consumer interaction, showing the active set of designs (*black dots*) using Rule 3—Maximize performance/cost ratio, recorded when the target metric reached 25. Compared to Fig. 11.8, the pace of movement along the design trajectory is slower

11.7 Discussion

The results substantiate the argument that interactions between Consumers and Producers in the post-design phase are essential to the process of innovation. Producers need to combine several types of learning in order to shape the post-design outcomes and to feed into the pre-design and design activities to follow. First, they need to gain tacit knowledge to achieve improvements in performance/cost ratio. In a sense, the design phase only offers the promise and potential for a certain performance/cost goal to be achieved. Whether that promise and potential is realized depends on whether the Producer has sufficient persistence and follow-through to avoid excessive diversification, and also sufficient flexibility and openness not to get stuck in a narrow range if it precludes other, more promising paths. Second, the Producer must continually learn and adapt to what the marketplace defines as 'value'. Certainly, market research and testing can help here, but this learning mostly happens through engagement with the market and Consumers through post-design activities of selling, servicing, and customizing.

Consumers, too, engage in post-design 'work' of a sort. Through their shifting and emergent preferences and surprising discoveries of new functions and new significance, they reshape the value landscape both for Producers and also other Consumers. In future experiments, we expect to see patterns of reinforcing interactions between Producers and Consumers, but maybe also inhibiting interactions, too, maybe in surprising ways.

While the results in this paper are illustrative and suggestive, we believe they begin to show value of Computational Social Science methods in studying innovation processes that span social and technical domains and diverse social groups. The architecture of the simulation system demonstrates that it is feasible to build rich computational models to study the simultaneous influence of several factors at once, and across different social levels. This is especially important when we are studying how agents react to and even generate novelty. In particular, our agent architecture for Consumers includes a comparatively rich set of capabilities to model situated cognition, both with artifacts and also with other agents. Also, our artifact architecture for Product demonstrates that it is possible to design an environment that is rich in possibilities without requiring the designer to (necessarily) plan or explicitly design all those possibilities ahead of time. For example, we chose three utility dimensions for Products based on the characteristics of graphs (i.e. degree count, clustering coefficient, and longest span in an embedding). But the simulation system could be greatly enriched if the utility dimensions were open-ended and endogenously created and diffused by agents. This is an example of the benefits of the computational approach.

Another significant benefit of computational modeling is the ability to run both exploratory and controlled experiments that have demonstrable relevance to real-world settings. The results presented for Setting 1—Producer Acting in Isolation—are still exploratory at this stage. However we were able to identify three qualitative distinctions between design trajectories. In future work, we aim to quantify and measure these distinctions so that we can do more rigorous experiments involving design trajectories as dependent variables.

The results presented in Setting 2—Consumers Acting in Isolation—take us closer to controlled experiments and statistical hypothesis testing. We were able to identify conditions where our Consumers were and were not able to endogenously form clusters of values. Clustering has a significant effect on the diffusion and adoption of innovation, both in the form clusters of early adopters and also clusters of resistance to change.

Another contribution of our research is demonstrating how to measure and evaluate changes in value systems in the context of innovation, both at the level of an individual, in a group, and in a population. In this paper we have presented and discussed three different analysis and representation methods—(1) design trajectories in the space of possible designs, (2) Value Space for populations of Consumers and their ideal product vectors; and (3) Utility Space for populations of Consumers (and Producers) to monitor and measure how their utility function changes over time. One of our most significant lessons from this research so far is the value of multiple simultaneous measurements and representations, because value systems is inherently multi-level, multi-dimensional, and even pluralistic, even within an individual agent.

With further experiments and results, we expect that our experiments will reveal emergent patterns of organization that shape the Producer's design choices. In particular, we believe that experiments will reveal self-reinforcing processes (both direct and indirect) where early Consumer learning and preference formation influences Consumer receptivity to new Products that have similar surface characteristics or offer similar performance dimensions, and this in turn influences

Producer decisions on the trajectory of new product introduction. We suspect that this behavior will appear at critical values of parameters that govern the rate of learning and adaptation by Producers and Consumers.

Acknowledgements This research is based upon work supported by the US National Science Foundation under Grant Nos. SBE-0915482, CNS-0745390, CMMI-116715 and CMMI-1400466. Any opinions, findings, and conclusions or recommendations expressed in this material are those of the author and do not necessarily reflect the views of the National Science Foundation.

References

Akintoye A, Beck M (2009) Policy, management and finance for public-private partnerships. Wiley, Oxford

Archibugi D, Howells J, Michie J (1999) Innovation policy in a global economy. Cambridge University Press, New York

Asimov M (1962) Introduction to design. Prentice-Hall, Englewood Cliffs

Auerswald P, Kauffman S, Lobo J, Shell K (2000) The production recipes approach to modeling technological innovation: an application to learning by doing. J Econ Dynam Contr 24(3):389–450

Avermaete T, Morgan EJ, Viaene J, Pitts E, Crawford N, Mahon D (2003) Regional patterns of innovation: case study of small food firms. In: DRUID summer conference 2003 on creating, sharing and transferring knowledge. Copenhagen, Denmark, June 12–14, 2003

Bower JL, Christensen CM (1996) Disruptive technologies: catching the wave. J Prod Innovat Manag 13(1):75–76

Castelfranchi C (2001) The theory of social functions: challenges for computational social science and multi-agent learning. Cognit Syst Res 2(1):5–38

Casti JL (1999) The computer as a laboratory. Complexity 4(5):12–14

Clancey WJ (1997) Situated cognition: on human knowledge and computer representations. Cambridge University Press, Cambridge

Cooke P (2001) Regional innovation systems, clusters, and the knowledge economy. Ind Corp Change 10(4):945–974

Danneels E (2004) Disruptive technology reconsidered: a critique and research agenda. J Prod Innovat Manag 21(4):246–258

Dooley K (1997) A complex adaptive systems model of organization change. Nonlinear Dynam Psychol Life Sci 1(1):69–97

Dosi G (1982) Technological paradigms and technological trajectories: a suggested interpretation of the determinants and directions of technical change. Res Pol 11(3):147–162

Dosi G, Nelson RR (2010) Technical change and industrial dynamics as evolutionary processes. In: Handbook of the economics of innovation. Elsevier, Amsterdam, pp 51–127

Edquist C (2005) Systems of innovation: perspectives and challenges. In: Oxford handbook of innovation. Oxford University Press, Oxford, pp 181–208

Epstein JM (2012) Generative social science: studies in agent-based computational modeling. Princeton University Press, Princeton

Epstein JM, Axtell R (1996) Growing artificial societies: social science from the bottom up. Brookings Institution Press, Washington, DC

Feldman DH, Csikszentmihalyi M, Gardner H (1994) Changing the world: a framework for the study of creativity. Praeger, Westport

Ferber J (1999) Multi-agent systems: an introduction to distributed artificial intelligence. Addison-Wesley, Harlow

Finke RA (1996) Imagery, creativity, and emergent structure. Conscious Cognit 5(3):381–393

Friedkin NE, Johnsen EC (1999) Social influence networks and opinion change. Adv Group Processes 16(1):1–29
Gärdenfors P (2004) Conceptual spaces: the geometry of thought. MIT Press, Cambridge
Gero JS (1990) Design prototypes: a knowledge representation schema for design. AI Mag 11(4):26
Gero JS (1996) Creativity, emergence and evolution in design. Knowl Base Syst 9(7):435–448
Gero JS (2000) Computational models of innovative and creative design processes. Technol Forecast Soc Change 64(2–3):183–196
Gero JS (2008) Towards the foundations of a model of design thinking. DARPA Project BAA07-21
Gero JS, Damski JC (1997) A symbolic model for graphical emergence. Environ Plann B Plann Des 24(4):509–526
Gero JS, Kannengiesser U (2004) The situated function–behaviour–structure framework. Des Stud 25(4):373–391
Gero JS, Kannengiesser U (2009) Understanding innovation as a change of value systems. In: Proceedings of growth and development of computer aided innovation third IFIP WG 5.4 working conference, CAI 2009, pp 249–257, Harbin, 20–21 Aug 2009
Gero J, Maher M (eds) (1993) Modeling creativity and knowledge-based design. Psychology Press, Florence
Gilbert N (2002) Varieties of emergence. In: Proceedings of agent 2002 conference: social agents: ecology, exchange, and evolution, pp 11–12, Chicago
Gilbert N, Conte R (1995) Artificial societies: the computer simulation of social life. Psychology Press, Florence
Goldstein J (1999) Emergence as a construct: history and issues. Emergence 1(1):49–72
Grace K, Saunders R, Gero J (2008) A computational model for constructing novel associations. In: Proceedings of the international joint workshop on computational creativity. Madrid, Spain, September 17–19, 2008
Holland JH (2000) Emergence: from chaos to order. Oxford University Press, New York
Hybs I, Gero JS (1992) An evolutionary process model of design. Des Stud 13(3):273–290
Kauffman S (1996) At home in the universe: the search for the laws of self-organization and complexity. Oxford University Press, Oxford
Kuhn TS (1996) The structure of scientific revolutions. University of Chicago Press, Chicago
Leydesdorff L (2000) The triple helix: an evolutionary model of innovations. Res Pol 29(2):243–255
Liggett RE (2010) Multivariate approaches for relating consumer preference to sensory characteristics. Ph.D. dissertation, The Ohio State University
Llerena P (2006) Innovation policy in a knowledge-based economy: theory and practice. Springer, Berlin
Macy MW, Willer R (2002) From factors to actors: computational sociology and agent-based modeling. Ann Rev Sociol 28:143–166
Malerba F, Nelson R, Orsenigo L, Winter S (2007) Demand, innovation, and the dynamics of market structure: the role of experimental users and diverse preferences. J Evol Econ 17(4):371–399
Miller JH, Page SE (2009) Complex adaptive systems: an introduction to computational models of social life: an introduction to computational models of social life. Princeton University Press, Princeton
Moldovan S, Goldenberg J (2004) Cellular automata modeling of resistance to innovations: effects and solutions. Technol Forecasting Soc Change 71(5):425–442
Pahl G, Beitz W, Feldhusen J, Grote KH (eds) (2007) Engineering design: a systematic approach. Springer, London
Repenning NP (2002) A simulation-based approach to understanding the dynamics of innovation implementation. Org Sci 13(2):109–127
Sahal D (1985) Technological guideposts and innovation avenues. Res Pol 14(2):61–82

Saunders R, Gero JS (2004) Situated design simulations using curious agents. AIEDAM 18(2):153–161

Saviotti P (1996) Technological evolution, variety, and the economy. Edward Elgar Pub, Aldershot and Lyme

Saviotti PP, Metcalfe JS (1984) A theoretical approach to the construction of technological output indicators. Res Pol 13(3):141–151

Sawyer RK, Sawyer RK (2012) Explaining creativity: the science of human innovation. Oxford University Press, Oxford

Scherer FM (1986) Innovation and growth: schumpeterian perspectives. MIT Press, Cambridge

Schumpeter JA (1942) Capitalism, socialism and democracy. Harper, New York

Smith GJ, Gero JS (2001) Emerging strategic knowledge while learning to interpret shapes. In: Gero JS, Hori K (eds) Strategic knowledge and concept formation III, Key Centre of Design Computing and Cognition. University of Sydney, Sydney, pp 69–86

Smith GJ, Gero JS (2005) What does an artificial design agent mean by being 'situated'? Des Stud 26(5):535–561

Sosa R, Gero JS (2004) A computational framework to investigate creativity and innovation in design. In: Gero JS (ed) Design Computing and Cognition '04. Springer, New York

Sosa R, Gero JS (2005) A computational study of creativity in design: the role of society. AI EDAM 19(4):229–244

Sosa R, Gero J (2007) Computational explorations of compatibility and innovation. In: León-Rovira N (ed) Trends in computer aided innovation, vol 250. IFIP, The International Federation for Information Processing, Springer, New York, pp 13–22

Taura T, Nagai Y (2005) Primitives and principles of synthesis process for creative design. In: Gero JS, Maher ML (eds) Preprints of international conference of computational and cognitive models of creative design VI, pp 177–194, Key Centre of Design Computing and Cognition Heron Island, Australia, 10–14 Dec 2005

Weiss G (2000) Multiagent systems. MIT Press, Cambridge

Wooldridge M (2008) An introduction to multiagent systems. Wiley, Chichester

Chapter 12
Modelling (pre-)Design Activities With a Multi-Stakeholder Perspective

Gaetano Cascini and Francesca Montagna

Abstract The paper proposes an extension of the Gero's Function-Behaviour-Structure (FBS) framework aimed at representing Need and Requirements and their relationships with the Function, the Behaviour and the Structure of an artefact. Needs and Requirements can be modelled as further types of variables to describe with the same formal approach of the situated FBS model the transformation processes which occur in the earlier stages of design, when the requirements still need to be specified. Furthermore the external world where needs are situated is split into the complementary perspectives of the different stakeholders influencing the adoption process of a new product, i.e. into buyers, users, beneficiaries and other outsiders. The extended model aims at supporting a more careful and detailed investigation of the processes that occur in the earliest stages of design, and specifically what happens in new product development activities. As carefully discussed in the introduction of the paper, such a shift in the designer's perspective appears as a crucial step to build an efficient design methodology for innovation.

Keywords Design for innovation • FBS framework • Multi-stakeholders analysis • Needs identification

12.1 Introduction

The design of innovative and successful products certainly requires a good deal of creativity by the people involved individually, or as a team, in the design process from the concept generation to the product embodiment and even further. In addition, design creativity by itself cannot be considered a sufficient factor to ensure the commercial success of a new product or, in a broader sense, its adoption by a certain community. Roughly speaking, the adoption process occurs when the

---

G. Cascini (✉)
Politecnico di Milano, Dip. di Meccanica, Via La Masa 1, 20156 Milan, Italy
e-mail: gaetano.cascini@polimi.it

F. Montagna
Politecnico di Torino, Dip. di Ingegneria Gestionale e della Produzione,
C.so Duca degli Abruzzi 24, 10129 Torino, Italy
e-mail: francesca.montagna@polito.it

© Springer Japan 2015
T. Taura (ed.), *Principia Designae – Pre-Design, Design, and Post-Design*,
DOI 10.1007/978-4-431-54403-6_12

economic actor proposing a new product on the market is able to cover his costs by selling a given volume of it at a given price and that, on the side of demand, a number of economic actors are willing to buy that volume of products at that price, and still gain some utility out of it (Cantamessa et al. 2012).

Indeed, the so-called Fuzzy Front End of New Product Development, which concerns the stages spanning from the opportunity identification to the concept definition, is characterized by high market and technological uncertainties. While the latter have been traditionally addressed by a multitude of studies both by academia and industry, just recently engineering design research has approached the proposition of methods and tools suitable to systematically support needs identification and concept ideation, so as to reduce market uncertainties. Still, the lack of structured means for supporting decision making within the front end of innovation (Appio et al. 2011) determines a huge waste of resources for developing products that in most cases turn out to be market failures (Stevens and Burley 1997; Borgianni et al. 2013).

In this context, the product user has become the focus of designers' attention, as witnessed by the success of research topics such as "co-design" and "user-centered design". This is certainly a necessary, but not sufficient step ahead towards the definition of design methods and tools suitable to guide the ideation of new products likely to be adopted, or in brief towards a design methodology for innovation. In fact, several stakeholders can play a significant role on the adoption process of a new product. As described in detail in the next chapter, it is more and more necessary to identify these mutual influences so as to properly recognize beforehand the motives and the values behind the success of a new product.

All in all, it means that design has to explicitly embrace all the activities related to the identification of the needs and the demands that people have or can be induced to have, as well as the study of the mechanisms behind the purchase and the adoption decision. Beyond the seminal tools and methods already available to support these activities, mostly derived from industrial practice, it is time to investigate with more consistent scientific means what works, what is appropriate, what produces real value, in turn what reliably brings to innovation.

In this perspective, this paper presents a model, built as an extension of the situated Function-Behaviour-Structure (FBS) framework (Gero and Kannengiesser 2004), suitable to explicitly represent any cognitive process occurring in the early stages of a New Product Development activity, or in a Pre-Design activity according to the terminology of this collection of papers. The model aims at being a reference to recognize and study best practices in industry, to identify strengths and weaknesses of a design methodology for what concerns its support to conceptual design, and to codify individual and collective thinking paths within innovation tasks.

The paper is structured as follows: the next section analyses with further details the motivation of this study, as well its related literature. The third section describes the original situated FBS framework, highlighting the reasons to use it as a reference, as well as what needs to be integrated for the purpose of this work. Then the authors' proposal for extending the FBS model is presented, with a double focus: from the one hand the representation of needs and their translation into a design specification, from the other hand the representation of the different stakeholders influencing the adoption process. The model is illustrated and discussed with a simple example aiming at clarifying its features and potentialities.

12.2 Building the Specification of a New Product

As introduced above, the identification and the fulfilment of customers' needs represent a critical issue for designing and developing successful products. The stream of research dedicated to user-centred design has dedicated major efforts to understand customer needs through the establishment of a closer relationship between designers and users. In most cases, needs identification is triggered by marketing inputs, and achieved through the observation of customer's behaviour so as to elicit the so-called Voice-of-the-Customer. Indeed, this approach proves to be very effective in order to improve existing products, but turns out to be quite limited when radical innovation is searched. In fact, it is largely recognized that "customers don't know what they want in the future" (Eisingerich et al. 2010), although competition demands companies to be "the first to give it to them". Therefore, despite its intrinsic limitations, the involvement of users in the design process has been growing for two decades since its first theorization (von Hippel 1986), and still it is an important trend in design (Borgianni et al. 2013). The currently most debated topics in this domain span from the proposition of practical tools to individuate seeded needs (Ward 2009) to the correlation of the emerging needs with the design specification (Ericson et al. 2009).

Nevertheless, two main limitations appear, one related to the object of study, the other to the means to conduct the study itself, as further discussed in the following two subsections. First, the user is not necessarily the most important actor influencing the purchase decision, and in most cases, he is not the only one. Second, while design research has matured appropriate means to study in detail the cognitive processes occurring in the design process once that the requirements are explicitly defined, no reference models exist to investigate with equivalent rigour the previous phases of product development.

12.2.1 A Multi-Stakeholder Perspective for Design

As anticipated before, in most instances, the innovation process is considerably more complicated than simply making sure that a single user/buyer and a seller will find mutual benefit from a transaction, so that the former will buy the product from the latter. In fact, after the purchasing decision, the product must be actually put to use (i.e., adopted) in order to deliver its benefits, and this process may also necessitate an investment of time and effort to learn how to use it, to set it up, etc.

Furthermore, the sale of the expected volume of products is not instantaneous, but typically follows a diffusion process (Mahajan et al. 2000). In this further process, the actors who have not adopted the innovative product yet are usually influenced by those who have successfully done so, either because of direct "word of mouth", or because of simple observation of the benefits.

It is even more important to notice that products and services seldom involve the interaction with a single actor (Laffont 1988; Cantamessa 2011), while more likely a multi-actor perspective is appropriate. While buyers and users are not necessarily the same person, the actor(s) that will ultimately benefit from the product might be different from either the buyer or the user. Indeed, also other stakeholders, such as installers or vendors, can influence the buyers. In some cases, the relationship between user and beneficiary is relatively straightforward to identify. In other cases, the transaction between a buyer and a seller impacts either positively or negatively someone who is external (i.e., an outsider not taking part) to the transaction. In economics, this is referred as externality (Laffont 1988). Cantamessa et al. (2012) illustrate this classification of stakeholders through a simple example: in the case of buses for public transport, the buyer is the purchasing office of the transport authority, the users are the bus drivers (but one might also include maintenance crews in this category) and the direct beneficiaries are the passengers. However, there are also "outsider" beneficiaries involved, such as citizens being affected by the emissions of the vehicle, or passers-by that may be impacted by the degree of pedestrian impact performance of the bus.

In turn, to have a successful innovation (i.e., to actually make commercial impact), the design of an innovative product or service should take into account all the phases that make up the innovation process and the specific decisions taken by the actors who take part in each phase. Failing to do so, i.e., neglecting any of these phases, can actually kill the acceptance of a new product even with good potential.

The concept of "Design for Innovation" can therefore be separated in at least four core components, namely "Design for Purchase" (by buyers), "Design for Adoption" (by users), "Design for Impact" (on the beneficiary) and "Design for Externalities" (on outsiders). Moreover, the study of these complex interactions implies taking into account also that actors involved do not act in isolation from the others. Therefore, Design for Innovation does not only require understanding the individual perspective of each actor, but also the influences that are reciprocally cast among the actors and—potentially—the conflicts occurring between their needs.

So far, no design methodologies exist to explicitly address innovation with such a perspective. A first move in this direction has been proposed by the authors in Cantamessa et al. (2012). Nor there exists any reference model to codify and analyse design activities devoted to the interpretation of the different stakeholders' needs, their mutual interactions, as well as their interpretation to build the design specification.

This paper summarizes the studies carried out so far by the authors with the objective of building an extension of the FBS framework suitable to fill this gap in the design research domain.

12.2.2 Needs and Requirements, Two Different Things

Customer needs are largely debated in the marketing literature since decades. A long-lasting stream of research deals with the "Voice of the Customer" (Griffin and Hauser 1993), especially with the focus on customer behaviour (Belch and Belch 2004). Besides the marketing literature, other authors from different fields claim the importance of including the identification of customer and market needs within new product development key activities.

Customer Needs can be categorised according to their urgency (e.g., from physiological necessities to means for achieving personal satisfaction) or according to their universality (e.g., relevant for the whole human being or highly based on individual judgment); however, in any case, these needs are the basic motivation for pushing people to change their situation (Beatty et al. 1985; Kahle et al. 1986; Maslow 1987).

In engineering design, marketing-oriented issues are considered success factors of product development projects (e.g., (Cooper and Kleinschmidt 1990; Calantone et al. 1993; Balachandra and Friar 1997; Ulrich and Eppinger 2008)). Similarly, in architecture and industrial design, the issue of creating a social design process has been assimilated since Alexander's seminal contributions in 1964 (Alexander 1964), up to the creation of social artefacts, i.e., aiding larger communities and supporting society (Krippendorff 2006). The concept of need has been also broadened from the functional to the emotional sphere of user experience (Ortiz Nicolas and Aurisicchio 2011; Pucillo and Cascini 2014), also through the virtualisation of interaction tests (Bordegoni et al. 2011). Also in industry, topics such as "co-design" or "user-centered design" are evidence of the importance of considering the user's perspective in design.

Some scholars, as well as some industrial practices, follow a different approach to the ideation of new products, not based on the observation of customer behaviour, but rather inspired by codified patterns of evolution of technical systems (Cascini 2012). Nevertheless, also in this case the concept of the new product has to be positioned with respect to the clusters of consumers the designers intend to address, i.e., to the needs they intend to satisfy. In other terms, needs can be derived from customers' inputs (both explicit and tacit) or postulated by the designer according to his/her experience and his/her consolidated know-how in that domain. The former category of needs involves explicit requests by the customers, as well as features that can be identified through the observation of users' behaviour with existing artefacts. Additionally, the needs postulated by the designer are more typically related to technology-driven artefacts and follow typical evolutionary patterns of products. The ability to identify tacit needs or to envision future and further needs is sometimes induced by new products and it is sufficient to generate an innovation process and to determine the success of a novel product. For example, Velcro or Post-it are typical outcomes of serendipity, but they are also the result of innovators' skill of creatively meeting the observation of reality with tacit customer needs. Conversely, Apple products are the fruit of technology-driven processes in which innovation has relied on the talent of anticipating the emerging needs induced by new technical products.

Despite the growing interest on needs (either explicit or tacit) as a compass for the development of innovative products, traditionally a design activity starts from a design specification to fulfil, which is related to, but not coincident with the needs intended to satisfy. In detail, the design specification is a "structured and formalised information about a product" (Ericson et al. 2009) consisting of a set of requirements, each entailing "a metric and a value". A proper design specification is therefore constituted by a list of measurable features such that it is possible to assess whether the needs are satisfied in a given context and constitute the reference for the development of a not-yet-designed product.

An accurate analysis of the differences between needs and requirements appears in (Ericson et al. 2009), where Ericson et al. distinguish between the customer's context, where values and needs are perceived by the users, and a second context, which is the product developers' context, where requirements and specifications are designed by the development team. The bridge is represented by the "Need Representation" stage in which needs are generated and designed by the heterogeneous development team on the basis of what has been found in the customer's context. During this process the information captured from the customer's context should be converted by the designer into information usable for the product development.

Nevertheless, in the engineering design literature, a clear distinction between methods that address Needs Identification or Requirements Definition does not exist, presumably because of an unclear distinction between the two concepts. Indeed, it is only possible to recognise an orientation towards one of the two targets. Beyond the debate on needs overviewed in the previous subsection, the literature on requirements specification has become much larger since the 1990s following the development of complementary approaches including Quality Function Deployment (e.g., (Clausing 1998)) and detailed prescriptive criteria to codify the information gathered from external sources (e.g., (Cooper et al. 1998)). A comprehensive survey about methods for mapping design requirements has been presented in Darlington and Culley (2002).

12.3 Building on the Situated FBS Model

Many models have been proposed in literature to represent the design process; some of them are prescriptive in nature, i.e., they aim at guiding design tasks; others are essentially descriptive, thus they represent designers' behaviour and thinking processes. An extensive survey of these models is proposed in Wynn and Clarkson (2005).

Among the applications of the descriptive models, it is worth mentioning their use within design protocol analyses as a means to analyse the design activities of individuals and teams, so as to identify typical patterns and behaviours (e.g., (Bierhals et al. 2007)), as well as to evaluate the impact of training activities (as in (Gero et al. 2012)). Besides, as discussed above, none of those models allows mapping, and thus studying, what concerns needs identification and the definition of the design specification.

With the aim of addressing such a gap, the authors decided to build on a well-acknowledged existing model, rather than creating a dedicated new one. This choice has several motivations: first, an extension of a well-known model is expected to be easier to understand and to be adopted by other scholars. Second, it allows leveraging the advantageous research experiences conducted so far by the design community. Eventually, an extended model, rather than a dedicated one, is intrinsically suitable for analysing the entire design process, and not just the earliest stages of new product development, therefore with a much wider spectrum of potential exploitations.

An acknowledged model for representing design cognitive processes is the situated FBS framework (Gero and Kannengiesser 2004), which has all the features to be extended according to the proposed objectives. Besides, since its first formulation in 1990 (Gero 1990), the FBS framework has been already evolved both by his main author and by other scholars, so as to extend its applicability to diverse contexts.

The FBS model considers three classes of variables as the object of the design activities:

- Functions (F), which describe the aim of the object, i.e., what the object is for;
- Structures (S), which describe the object components and their relationships, i.e., what the object is; and
- Behaviours (B), which describe the attributes that are derived or expected to result from the structure (S) variables of the object, i.e., what the object does.

These variables undergo eight reference processes (Fig. 12.1, left); five of them convert the posited functions sequentially into design descriptions. The first is called the *formulation* step and transforms functions F into a description of behaviour Be of an artefact that is expected to perform the previous functions. Then, the expected behaviour is transformed by a *synthesis* step into a structure S of the artefact by which it may show its behaviour Be. Subsequently, in a third step, called *analysis*, the actual behaviour Bs of the artefact with this structure S is derived. Fourthly, this actual behaviour is evaluated by comparing it with the expected behaviour. If this *evaluation* is satisfactory, a design description D is *documented* for manufacturing the artefact with the structure S. If the evaluation is not satisfactory, the design process returns to previous steps, defining three elementary loop-back stages and defining the design process as an iterative procedure.

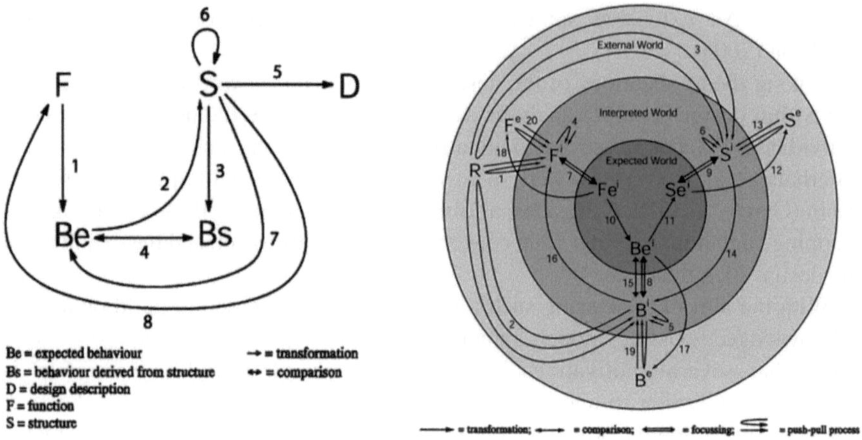

Fig. 12.1 The original FBD framework (*left*, (Gero 1990)) and the situated one (*right*, (Gero and Kannengiesser 2004))

Moreover, the situated model introduces three different types of environments, recursively linked together, in which those processes can take place:

- The *external world* is made of representations outside the designer.
- The *interpreted world* is made of sensory experiences, concepts and interpreted representations of that world with which the designer interacts.
- The *expected world* is the world in which the effects of the actions of the designer are imagined according to the current goals and the interpretations of the present state of the world.

The situated model is shown in Fig. 12.1 (right), where the three different worlds, the variable types, and their 20 characteristic transformation processes are depicted.

An immediate consideration that clearly emerges from the analysis of the FBS framework, with respect to the scope of the present work, is that needs identification as well as requirements definition are not fully represented.

In actuality, Gero and Kannengiesser (2004) explicitly refer to the requirements (R) of a design problem, but the description of the formulation process is limited to the statement "the design agent interprets the explicit requirements (R) by producing the interpreted representations F^i and, eventually, B^i and S^i". Compared with the careful description of the following design processes, the requirements definition appears too simplistic, probably due to the traditionally limited relevance assigned in design theory to user needs recognition.

Moreover, an analogous comment appears also in Vermaas and Dorst (2007), despite the purpose of the paper is different from the present work: it actually claims that "designing starts with a client's intentional aim or desire, and produces a physicochemical description of an artefact by which the client can make the aim or desire come true", thus highlighting that the design process covers a more extended range than a translations of some requirements into a functional specification. In other terms, the first task accomplished a design agent, i.e., identification of user's intentional aims and desires is not adequately represented into the FBS design framework.

These considerations support the authors' intention to integrate, within the FBS framework, an explicit representation of Needs and Requirements to have proper means to model the entire product development process from the earliest stages using the same formalism already proposed by Gero and Kannengiesser. Starting from these observations, the authors have formulated a proposal for an extended FBS model, characterized by two further types of variables and their related processes, as detailed in the following section.

12.4 The Extended FBS Model

According to the objectives of the present work and to the limitations of the FBS framework discussed in the previous section, the authors have proposed the integration of two further explicit classes of variables (Cascini et al. 2013).

- Needs (N): an expression of a perceived undesirable situation to be avoided or a desirable situation to be attained. This situation can be perceived by any of the actors involved in the product life from the purchasing phase to each stage of use and disposal. Needs can be explicitly stated to the designer or perceived by the designer because of being extracted (or even postulated) by the observation of users' behaviour.
- Requirements (R): a measurable property related to one or more Needs. They "are structured and formalised information about a product" and "consist of a metric and a value".

These two new sets of variables intend to investigate, as in the proposal of Vermaas and Dorst (2007), the concept "what it is for". This expression in Gero and Kannengiesser (2004) is considered sufficient to define Function, but as Vermaas and Dorst claim, everyday descriptions of the use of artefacts distinguish between purposes and functions. "Lasers are nowadays used as pointers during slide presentations. If such use is described, one can say that the laser's purpose is to highlight locations on projection screens, whereas its function is still to produce light. Then, purpose and function indeed refer to two different things, which becomes manifest when the use of the pointer fails, say because there is too much light falling on the projection screen: assuming that the laser still emits its regular amount of light, one can then say that the laser fails to achieve its purpose but not its function" (Vermaas and Dorst 2007).

The Formulation phase (the interpretation of the requirements and the related transformation into functions, behaviours and structures), in the design process, requires formalisation in addition to interpretation. The original FBS model considers the activity of interpretation, but does not take into account the formalisation effort. Only formalisation can convert interpretations into formalised requirements descriptions from needs.

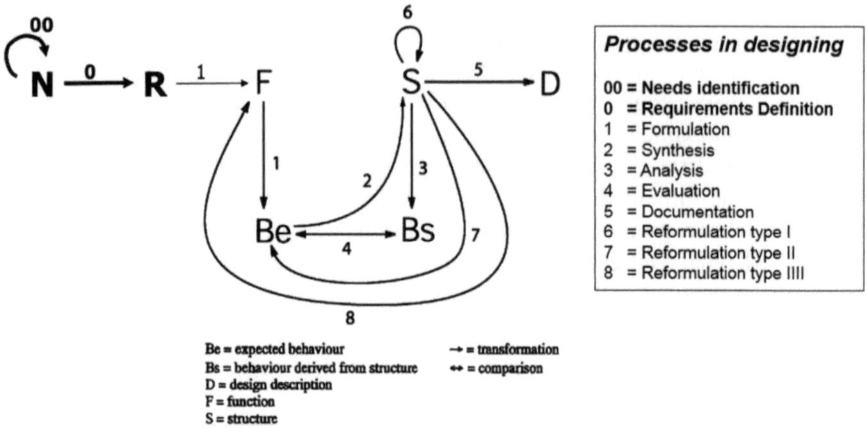

Fig. 12.2 A schematic representation of the ten main processes of designing in the Extended FBS model. The *bold symbols* refer to the variables and processes proposed by the authors

By introducing Needs and Requirements, the Formulation phase proposed in the original FBS framework must be reviewed in terms of those processes that definitively substitute the direct processes from the external word of R variables to the interpreted word of F^i, S^i, and B^i variables (e.g., Processes 1, 2, 3 and 18 in Fig. 12.1, right), which leads to the definition of two further phases: Needs Identification and Requirements Definition.

Overall, the design process thus consists of ten, rather than eight main processes, as depicted in Fig. 12.2.

It is interesting to notice that Gero's classification in terms of an external world, an interpreted world and an expected world still remains useful to "situate" Needs and Requirements as well. The new classes of variables, N and R, are subjected to the same transitions proposed in the original FBS model. This means that the extended FBS situates the new variables in the three worlds of the FBS framework, which produces the following: N^e, N^i, Ne^i, R^e, R^i and Re^i.

With respect to the above-mentioned work by Ericson et al. (2009) where customers and developers' contexts are dealt with separately, the adoption of Gero and Kannengiesser's "situatedness" allows for explicit representation of the links among these different worlds. Therefore, the proposed extension of the FBS framework allows for the description of processes such as those take place in "co-design" and "user-centered design" activities, which could not properly be modelled in the original framework.

12.4.1 Needs Identification

Needs identification is one of the two major steps that enrich the original situated FBS framework. As shown in Fig. 12.3a, the proposed extended model describes Needs Identification as constituted by three elementary processes:

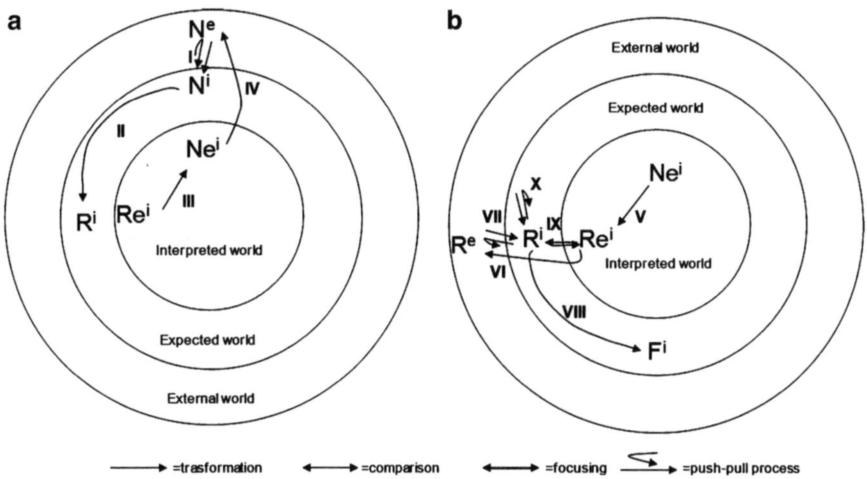

Fig. 12.3 Extended FBS model (Cascini et al. 2013): Needs Identification (**a**) and Requirements Definition (**b**)

- Process I investigates customer needs N^e, thus producing N^i variables (interpretation).
- Process II transforms N^i into R^i variables (transformation). These R^i are a preliminary set of requirements, useful to better categorize the gathered needs.
- Process III transforms the initial expected requirements Re^i into Ne^i variables (transformation). This step ensures that needs not provided by customers, but necessary, have the chance to be taken into consideration.
- Process IV transforms Ne^i into N^e variables (transformation) to validate the expected requirements with the customers. In case of negative feedback, the emerging external needs N^e can be analysed and interpreted to reformulate the requirements (through processes I and II).

In the proposed Needs Identification step, processes I and II correspond to analysis, while III and IV correspond to synthesis. This means, in general, that Needs Identification consists initially in an analysis process that creates a mental link between the captured Needs and possible Requirements in a sort of feasibility analysis. The process of synthesis produces a list of needs that can be deducted by the expectations of the designer (in the form of Requirements) and can be validated by the customer. Therefore, according to these alternative transitions, the identified Ne can be directly provided by the customer and/or can be anticipated by the designer who proposes the list of needs description to the customer.

12.4.2 Requirements Definition

Once that the Needs intended to fulfil have been identified, the design process moves to the definition of the design specification, i.e., the list of requirements to be satisfied.

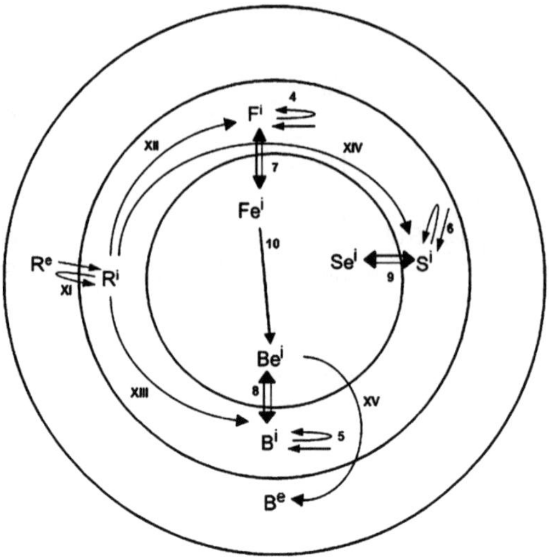

Fig. 12.4 Extended FBS model: the revisited formulation step

The Requirements Definition process (Fig. 12.3b), thus, starts from the complete list of expected needs Ne^i previously identified and it is made up of:

- Process V transforms the initial expected needs Ne^i into a first complete set of Re^i (transformation).
- Process VI expands the Re^i set into a bigger or equal number of R^e variables (transformation).
- Process VII uses R^e and results in their interpretation R^i, often through the constructive memory.
- Process VIII transforms the subset of R^i implying an active role of the product, i.e., not related to design constraints, into F^i variables (transformation).
- Process IX focuses on a subset ($Re^i \subseteq R^i$) of R^i to generate an initial requirement state space (focusing).
- Process X uses constructive memory to derive further R^i.

12.4.3 *Formulation*

Through the integration of Needs identification and Requirements definition into the Gero's FBS model, the Formulation step has to be updated at least for those parts that involve Requirements. Changes, as shown in Fig. 12.4, consist in those processes that definitively substitute the direct processes from the external word of R variables to the interpreted word of F^i, S^i, and B^i variables:

- Process XI reuses R^e to obtain definitive R^i (interpretation). This step is oriented to the creation of definitively validated interpreted Requirements that can be correctly elaborated.

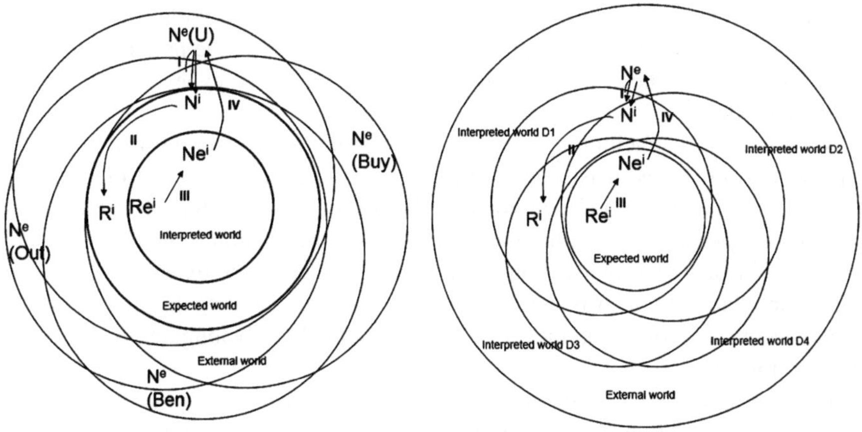

Fig. 12.5 Multi-actor perspective of the extended FBS framework: the four stakeholders of the adoption process (*left*); the different interpretations of a design team (*right*)

- Processes XII, XIII, XIV transform R^i into F^i, S^i and B^i variables, respectively (transformation). These F^i are not a preliminary set of functions anymore, but constitute, together with the other interpreted variables S^i and B^i, a detailed comprehensive set of design variables.
- The other processes from the 4th to the 10th are kept as in the original model.

12.4.4 *Situating Needs and Requirements in a Multi-Actor Perspective*

As discussed in Sect. 12.2.1, whatever is the object of design it is necessary to take into consideration the different perspectives of the diverse actors that share a certain degree of relationship with it. Each stakeholder has his own needs, which do not necessarily overlap with each other. Therefore, in the logic of the FBS framework, it means that different external worlds have be taken into account, each representing the domain where to situate the different actors' needs.

According to the classification of the different stakeholders into four categories Buyers (Buy), Users (U), Beneficiaries (Ben) and Outsiders (Out) proposed in Cantamessa et al. (2012), the situated FBS model should then generally entail four partially overlapping external worlds as depicted in Fig. 12.5 (left).

Besides, Interpreted and Expected worlds are proper of the designer mind: they are several if diverse designers are considered, but for each of them there is of course just one interpreted and one expected world. When considering a design team, according to the specific objective of the study, one can merge the potentially different interpreted and expected worlds of the different designers assuming that, by working together, they do share their expectations and interpretations. The last statement is reasonably acceptable at least for the expected world, once that the members of a design team have agreed on a shared vision on the project outcomes. Figure 12.5 (right) shows a scheme representing the multi-designers perspective of the FBS model.

With the aim of facilitating the readability of the figure and most of all of the represented processes, it is suggested to "open" the diagram, so as to reduce the apparent overlap of the different domains. For example, still referring to the multi-stakeholder perspective of the adoption process, the diagrams in Fig. 12.6 are obtained. Clearly, the intersections between the circles of the external worlds can change in relation to the way with which needs are shared among different actors. If some needs are common, they can be represented as belonging to the same intersection set.

By looking at Fig. 12.6 that represents the Need Identification (above) and the Requirement definition phases (below) respectively, one can deeply analyse which processes emerge when a designer considers needs by different actors. Specifically, in the Need Identification phase, the process I where customer needs N^e are investigated and the process IV that validates the expected requirements with the customers split into the four different categories of actors. This modification leads to consider one interpretation process for each actor and four further processes eventually appear:

- Process I (Buy/U/Ben/Out): buyer/use/beneficiary/outsider needs N^e(Buy/U/Ben/Out) are investigated, thus producing N^i variables.
- Process IV (Buy/U/Ben/Out) transforms Ne^i into N^e(Buy/U/Ben/Out) variables to validate the expected requirements with the buyer/use/beneficiary/outsider.

Furthermore, Processes II and III are situated in worlds that are unique, and hence remain the same. Similarly, in the Requirement Definition phase, the processes V, VIII, IX, X do not withstand modifications, while the processes VI e VII are distinguished for each actor. Given j = buyer, user, beneficiary, outsider.

- Process VI expands the Re^i set into a bigger or equal number of R^e(j) variables.
- Process VII uses R^e(j) and results in their interpretation R^i.

12.4.5 Potential Applications of the Extended Multi-Stakeholder FBS Model

As for the original situated FBS framework, the extended model developed by the authors and described in this paper allows for multiple applications, from the "simple" illustration of the multi-faced focus of a design activity, to the detailed analysis of a design method or the behaviour of a designer.

An example application of the extended FBS framework to review and improve an existing design methodology focused on user-device interaction (IDIM) has been published in Filippi et al. (2013). The IDIM methodology has been first

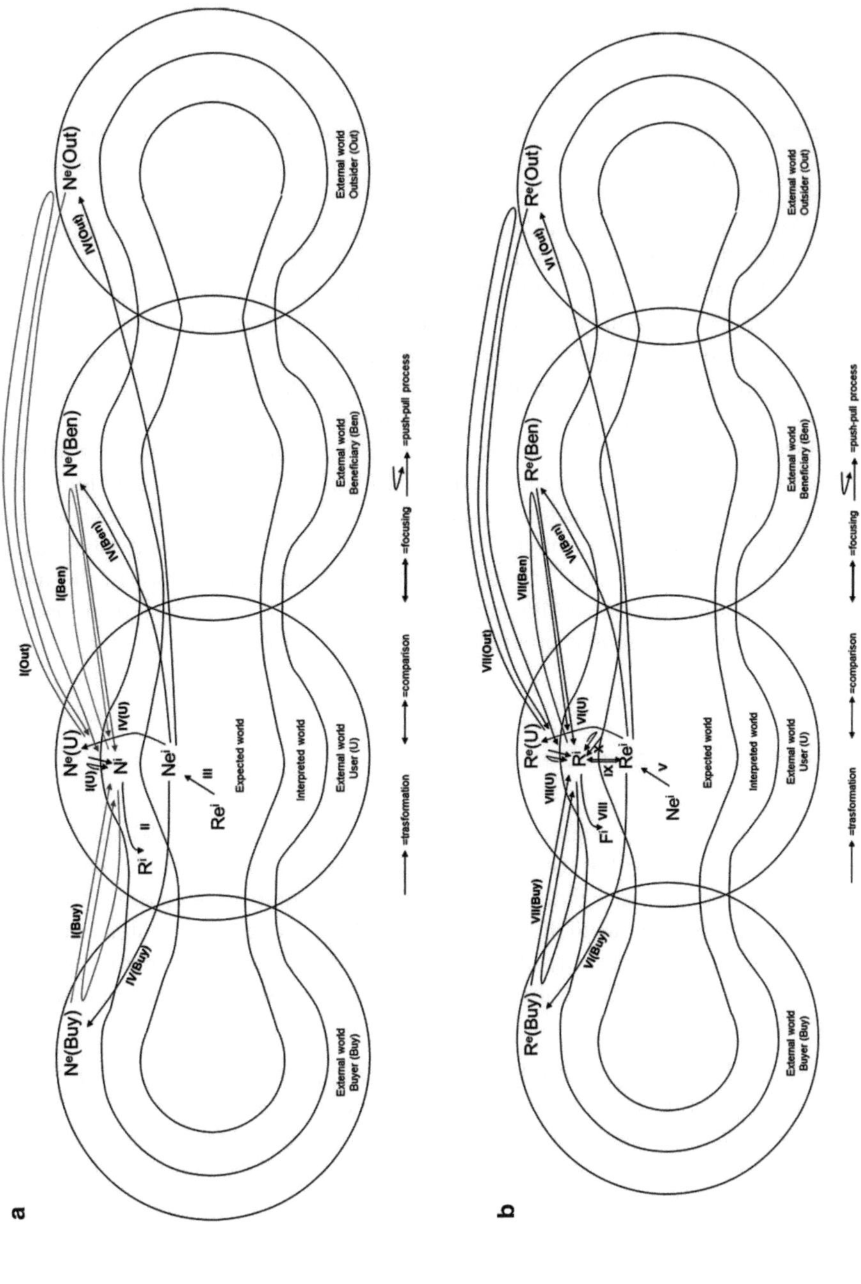

Fig. 12.6 A multi-stakeholder perspective of the preliminary design phases: **(a)** Needs Identification; **(b)** Requirement definition

mapped against the extended FBS model, looking for correspondences and eventual discrepancies between its prescriptive procedure and the processes described in the previous section of this paper. Then, the collected discrepancies have been analysed in detail. As a result, several criticalities emerged, mostly related to a lack of support of IDIM for some analysis and decision steps that were substantially left to the intuition of the designer. Based on those results, a number of suggestions about possible improvements of the IDIM have been produced.

Besides, a more typical use of the extended model is the investigation of a design activity through a protocol analysis; in this case, the model is used as a reference to codify the object of the design discourse (in terms of the types of variables involved) and the related transformations (as classified by the different processes). Examples of this kind of applications have been published in Cascini et al. (2013), where the analysis deals with an innovation project with Whirlpool Europe concerning the reduction of water and energy consumption of a washing machine, and in Cascini and Montagna (2014) where the case study refers to a brainstorming session aiming at the definition of the specification for a new production line, within an Italian company operating in the aseptic filling sector.

Two reference tables can be used to guide the application of a coding scheme based on the extended FBS model: the first (Table 12.1) collects the elementary statements cast by each actor (e.g., in a co-design brainstorming session) and associates to each statement the involved variable(s). The second table (as in Table 12.2) recognizes the evolution of the thinking process through the transformations applied to the design variables, and hence classifies the type of process according to the definitions provided in the previous sections.

12.4.6 Illustrative Example

Just with the purpose of clarifying the logic of the proposed model, some hypothetical reflections within the design of a kettle are here analysed. For the sake of simplicity, the multi-stakeholder perspective is here omitted. In general, customers provide several unstructured external needs such as reduced heating time, no maintenance, transportability of the device, volume capacity in order to make tea for four people, etc. These different needs feed the Need Identification step, in particular:

- Process I: uses N^e provided by customer to produce N^i variables such as “avoid formation of deposit” or “avoid impairing the following usages”.
- Process II: transforms N^i into R^i variables such as “subsequent usages do not create variations of the boiling time, as well as in the chemical/physical or organoleptic features of water”.

Indeed, the designer progressively attributes a metric and a target value to each R^i; eventually (through the synthesis process V and/or by choosing a subset of R^i in the process IX) the expected requirements Re^i are expressed by means of

Table 12.1 Example Table for listing the variables identified through a protocol analysis of a design task involving the interpretation of the needs of several stakeholders (Buy = Buyer, U = User, Ben = Beneficiary, Out = Outsider, D = Designer), classified according to the proposed extended FBS model

Actor	Actor's statement	Associated variable
Buy	*Statement #1*	N^e(Buy)
D	*Statement #2*	N^i
U	*Statement #3*	N^e(U)
	...	
D	*Statement #N*	R^i

Table 12.2 Example Table for listing the processes recognized for the transformations of the variables of Table 12.1

Design phase	Actors involved	From	To	Process
Needs identification	Buy→D	*Statement #1*	*Statement #2*	I(Buy): N^e(Buy)→N^i
	U→D	*Statement #3*	...	I(U): N^e(U)→N^i
	D	...	*Statement #N*	II: N^i→R^i
Requirements definition	D→U			VI(U): Re^i→R^e(U)
	Out→D			VII(Out): R^e(Out)→R^i

measurable technical features of the device and they constitute design constraints. In this case, it is possible consider requirements such as "boiling time <3 min", "Δ water hardness < 10 mg/L after the use", "water without deposit (<0.2 mm) in the mug", etc. The other design processes concern:

- Process III: transforms the initial expected requirements Re^i into Ne^i variables such as "short time for preparing hot water", "invariant water taste and healthiness", "no visible deposit inside the poured water".
- Process IV: transforms Ne^i into Ne variables: the interpreted Needs are proposed to candidate users to verify their appeal.

The processes that constitute the Requirement Definition process consider the complete expected needs Ne^i previously identified (e.g., "short time for preparing hot water", "invariant water taste and healthiness", "no deposit visible inside the structure"):

- Process V: transforms the initial expected needs Ne^i into a complete set of Re^i (as "boiling time <3 min", "Δ water hardness <10 mg/L after the use", "deposit <0.2 mm in the mug", "volume capacity = 1 L", etc.).

Table 12.3 Example processes occurring in the early stage of design of a kettle, classified according to the proposed extended FBS model

Design phase	Process	From	To	Process
Needs identification	I: $Ne \rightarrow N^i$	No maintenance	Avoid impairing the following usages	I: $Ne \rightarrow N^i$
			Avoid formation of deposit	
	II: $N^i \rightarrow R^i$	Avoid impairing the following usages	Subsequent usages of the kettle do not create variations of the boiling time, as well as of the chemical/physical or organoleptic features of water	II: $N^i \rightarrow R^i$
	III: $Re^i \rightarrow Ne^i$	The hardness of water boiled in the kettle should not increase more than 5 mg/L, nor the poured water should contain particles bigger than 0.2 mm	Invariant water taste and healthiness	III: $Re^i \rightarrow Ne^i$
			No visible deposit inside the poured water	
	IV: $Ne^i \rightarrow Ne$	Invariant water taste and healthiness	Boiled water should not impact the taste of the tea/coffee	IV: $Ne^i \rightarrow Ne$
			Safety of boiled water after years of usage	

Requirements definition	V: $Ne^i \rightarrow Re^i$	Kettle capacity sufficient to prepare four cups of tea/coffee	Volume capacity = 1 L	V: $Ne^i \rightarrow Re^i$
	VI: $Re^i \rightarrow Re$	Volume capacity = 1 L	Volume capacity = 1 L	VI: $Re^i \rightarrow Re$
			Height of the kettle < 300 mm	
	VII: $Re \rightarrow R^i$	Height of the kettle < 300 mm	Height of the kettle < standard scaffolds height	VII: $Re \rightarrow R^i$
	VIII: $R^i \rightarrow F^i$	Volume capacity = 1 L	Kettle contain water	VIII: $R^i \rightarrow F^i$
	IX: $R^i \rightarrow Re^i$	Subsequent usages of the kettle do not create variations of the boiling time, as well as in the chemical/physical or organoleptic features of water	After 1,000 usages of the kettle the boiling time to heat 1 L of water should remain less than 3 min	IX: $R^i \rightarrow Re^i$
			The hardness of water boiled in the kettle should not increase more than 5 mg/L, nor the poured water should contain particles bigger than 0.2 mm	
FBS formulation	XI: $Re \rightarrow R^i$	Boiling time < 3 min	Heating time of 1 L of water from 15 to 100 °C < 3 min	XI: $Re \rightarrow R^i$
	XII: $R^i \rightarrow F^i$	Heating time of 1 L of water from 15 to 100 °C < 3 min	The kettle increases water temperature from 15 to 100 °C	XII: $R^i \rightarrow F^i$
	XIII: $R^i \rightarrow B^i$	Heating time of 1 L of water from 15 to 100 °C < 3 min	The kettle supplies through an electric resistance 356 kJ to water (by Joule effect) in less than 3 min	XIII: $R^i \rightarrow B^i$
	XIV: $R^i \rightarrow S^i$	Heating time of 1 L of water from 15 to 100 °C < 3 min	The kettle parts in contact with water must be made with materials capable of working at 100 °C	XIV: $R^i \rightarrow S^i$

- Process VI: transforms Re^i into Re variables. The Re set is a bigger set of variables. It can contain variables directly referable to a specific Re^i such as "Volume capacity = 1 L" or other variables not explicitly considered in Re^i, such as "height of the kettle <300 mm".
- Process VII: Re are interpreted to produce R^i variables, as for instance "Height of the kettle < standard scaffolds height".
- Process VIII: transforms R^i into F^i variables such as "increasing water temperature until boiling status", "contain water", etc.

In parallel, a process which concerns mainly the R^i set and which derives the subset R^i can be conducted and consequently processes XI, XII, XII and XIV definitively transform variables from the external word to the interpreted word of F^i, S^i, and B^i variables. The other processes from the 4th to the 10th are maintained as in the original model. Table 12.3 presents a partial list of exemplary processes of each of the above-mentioned transformations. It is worth noticing that the purpose of the table is not to provide an exhaustive specification for the design of a kettle, but just to clarify the meaning of the added and revised processes occurring according to the proposed extended FBS model.

12.5 Conclusions

The present paper aims at contributing to the growing scientific discussion about needs identification and requirements specification by introducing an explicit representation of these different elements of a design activity. Furthermore, it highlights the importance of considering the needs that a product aims at fulfilling in a multi-stakeholders perspective, i.e., by considering all the actors who are involved in the product/service life from the purchasing phase to each stage of use and disposal.

While many scholars debate about the importance of considering the multi-faceted aspects of needs, as well some attempts exist to define design specifications with multi-stakeholders lists of requirements, there are still no models supporting a holistic representation of needs and their mutual relationships. Neglecting the impact of multi-actor contexts and influence elements can actually kill the innovation process even in the case of products with good potential. On the contrary, mapping needs and formulating requirements with a robust and comprehensive approach can be essential for adoption and diffusion processes, being it related to a product or to a service, from mass production to made-to-order businesses.

Therefore, Needs Identification and Requirements Definition have to be considered as two essential and distinguished phases of a design activity, and needs have to be contextualised with respect to their different owners.

The paper proposes an extension of the situated FBS model by Gero and Kannengiesser that is suitable to address such a critical issue. Specifically, the extended model proposes the explicit representation of a new type of variable, namely the Needs, and a more articulated representation of Requirements, a different type of variable already appearing in some of Gero's publications.

The aim of the proposed model is to support a more careful and detailed investigation of the processes that occur in the earliest stages of design. Therefore, it might be considered globally prescriptive, in terms of a general recommendation to include in any design activity a clear identification of the Needs to be addressed and then a careful definition of the Requirements specification. These should occur before proceeding with the formulation of the Function, Behaviour and Structure design variables. Additionally, the proposed model has mostly a descriptive nature, as a general framework to describe actual product definition processes and also to map and analyse the outcomes of the new user-oriented design practices. An illustrative example about the design of a kettle clarifies the meaning of the added variables and processes and the degree of detail that the model allows to represent.

The model also allows mapping the reciprocal influences that the different stakeholders might have on each other. Beyond the explicit representation of variables describing Needs and Requirements, the extended model suggests the identification of several external worlds, each representing the context of a different stakeholder. These are schematically classified into four main categories, namely Buyer, User, Beneficiary and Outsider. By distinguishing the need and requirement variables related to these external worlds and the processes they are involved in, it is possible to study a design activity with a detailed representation of the critical factors of the product-planning phase. Specifically, the model is suitable to represent tasks such as the interpretation of stakeholders' needs, their translation into formal requirements, the proposition of trade-offs between conflicting requests and all the decisions occurring in the formulation stage of design.

In general, the proposed model is expected to be fruitfully applicable with several purposes, such as the analysis of the activity of an innovation team, the verification of the impact of a training on product planning, and the quality of a design method. From this perspective, the main limitation seems to be the time needed to carefully classify a design protocol, given the increased number of variable types, worlds and actors to be considered.

Finally, according to the authors' knowledge, no other models currently exist to conduct detailed investigations on the earliest phases of product development. Given the importance of the latters, especially within innovation activities, the model presented in this paper can be considered worthy of further investigation and improvement.

References

Alexander C (1964) Notes on the synthesis of form. Harvard University Press, Cambridge

Appio FP, Achiche S, McAloone T, Di Minin A (2011) Understanding managers decision making process for tools selection in the core front end of innovation. In: Proceedings of the

International Conference on Engineering Design (ICED11), vol 10. Copenhagen, August 15–19, pp 102–113
Balachandra R, Friar K (1997) Factors for success in R&D projects and new product innovation: a contextual framework. IEEE Trans Eng Manag 44(3):276–287
Beatty SE, Kahle LR, Homer P, Misra S (1985) Alternative measurement approaches to consumer values: the list of values and the Rokeach value survey. Psychol Market 2(3):181–200
Belch GE, Belch MA (2004) Advertising and promotion: an integrated marketing communications perspective, 6th edn. McGraw-Hill, New York
Bierhals R, Schuster I, Kohler P, Badke-Schaub PG (2007) Shared mental models – linking team cognition and performance. CoDesign Int J Cocreation Des Arts 3(1):75–94
Bordegoni M, Ferrise F, Lizaranzu J (2011) Use of interactive virtual prototypes to define product design specifications: a pilot study on consumer products. In: Proceedings of the IEEE – ISVRI, Singapore, March 19–23, doi:10.1109/ISVRI.2011.5759592.
Borgianni Y, Cascini G, Pucillo F, Rotini F (2013) Supporting product design by anticipating the success chances of new value profiles. Comput Ind 64(4):421–435
Calantone RJ, Di Benedetto CA, Divine R (1993) Organisational, technical and marketing antecedents for successful new product development. R&D Manag 23:337–351
Cantamessa M (2011) Design... but of what? In: Birkhofer H (ed) The future of design methodology. Springer, London
Cantamessa M, Cascini G, Montagna F (2012). Design for innovation. In: Proceedings of the design conference (DESIGN 2012), Dubrovnik, May 17–20, p 747–756
Cascini G (2012) TRIZ-based anticipatory design of future products and processes. J Integrated Des Process Sci 16(3):29–63
Cascini G, Montagna F (2014) Situating needs and requirements in a multi-stakeholder context. In: Proceedings of the sixth international conference on design computing and cognition (DCC14), June 23–25, Springer, London, pp 377–395
Cascini G, Fantoni G, Montagna F (2013) Situating needs and requirements in the FBS framework. Des Stud 34:636–662
Clausing D (1998) Total quality development, 4th edn. ASME, New York
Cooper RC, Kleinschmidt EJ (1990) New products: the key factors in success. American Marketing Association, USA
Cooper R, Wootton AB, Bruce M (1998) Requirements capture: theory and practice. Technovation 18(8–9):497–511
Darlington MJ, Culley SJ (2002) Current research in the engineering design requirement. Proc IME B J Eng Manufact 216(3):375–388
Eisingerich AB, Bell SJ, Tracey P (2010) How can clusters sustain performance? The role of network strength, network openness, and environmental uncertainty. Res Pol 39:239–253
Ericson Å, Müller P, Larsson T, Stark R (2009) Product-service systems: from customer needs to requirements in early development phases. In: Proceedings of the 1st CIRP industrial product-service systems (IPS2) conference, Cranfield University, UK, April 1–2, pp 62–68
Filippi S, Barattin D, Cascini G (2013) Analyzing the cognitive processes of an interaction design method using the FBS framework. In: Proceedings of the 19th international conference on engineering design (ICED13), Seoul, August 19–22
Gero JS (1990) Design prototypes: a knowledge representation schema for design. AI Mag 11 (4):26–36
Gero JS, Kannengiesser U (2004) The situated function-behaviour-structure framework. Des Stud 25(4):373–391
Gero JS, Jiang H, Williams CB (2012) Design cognition differences when using structured and unstructured concept generation creativity techniques. In: Proceedings of the 2nd international conference on design creativity (ICDC2012), Glasgow, September 18–20
Griffin AJ, Hauser JR (1993) The voice of the customer. Market Sci 12(1):1–27
Kahle LR, Beatty SE, Homer P (1986) Alternative measurement approaches to consumer values: the List of Values (LOV) and Values and Life Style (VALS). J Consum Res 13(3):405–409

Krippendorff K (2006) The semantic turn. A new foundation for design. Taylor & Francis, Boca Raton

Laffont JJ (1988) Fundamentals of public economics. MIT, Cambridge

Mahajan V, Muller E, Wind J (2000) New product diffusion models. Kluwer, New York

Maslow AH (1987) Motivation and personality, 3rd edn. Harper & Row Publishers, New York

Ortiz Nicolas JC, Aurisicchio M (2011) The scenario of user experience. In: Proceedings of the international conference on engineering design (ICED11), vol 7. Copenhagen, Denmark, August 15–19, pp 182–193

Pucillo F, Cascini G (2014) A framework for user experience, needs and affordances. Design Stud 35:160–179

Stevens G, Burley J (1997) 3000 Raw ideas=1 commercial success! Res Tech Manag 40(3):16–27

Ulrich KT, Eppinger SD (2008) Product design and development. McGraw-Hill, New York

Vermaas PE, Dorst K (2007) On the conceptual framework of John Gero's FBS-model and the prescriptive aims of design methodology. Des Stud 28:133–157

von Hippel E (1986) Lead users: a source of novel product concepts. Manag Sci 3(7):791–805

Ward D (2009) Needs seeded strategies. J Appl Econ Sci 4(9):441–456

Wynn D, Clarkson J (2005) Models of designing. In: Clarkson J, Eckert C (eds) Design process improvement – a review of current practice. Springer, London, pp 34–59

Chapter 13
A New Perspective for Risk Management: A Study of the Design of Generic Technology with a Matroid Model in C-K Theory

Pascal Le Masson, Benoit Weil, and Olga Kokshagina

Abstract Risk management today has its main roots in decision theory paradigm (Friedman and Savage, J Polit Econ 56:279–304, 1948). It consists in making the optimal choice between given possible decisions and probable states of nature. In this paper we extend this model to include a design capacity to deal with risk situations.

A design perspective leads to add a new action possibility in the model: to design a new alternative to deal with the probable states of nature. The new alternative design might also "create" new risks, so that a design perspective leads also to model the emergence of new risks as an exogenous "design process". Hence a design perspective raises two issues: can we design an alternative that would lower the risk? Does this new alternative create new risks?

We show (1) that minimizing known risks consists in designing an alternative whose success is independent from all the known risks—this alternative can be considered as a generic technology. We show (2) that the design of this generic technology depends on the structure of the unknown, ie the structure of the space generated by the concept of risk-free alternative. (3) We identify new strategies to deal with risks as dealing with the unknown.

Keywords Decision theory • Design theory • Evolutionary model • Generic technology design • Independence • Matroid • Risk management • Robustness • Structure of the unknown

Unknowledgements: The Chair of Design Theory and Methods for Innovation, French Council for Energy.

P. Le Masson (✉) • B. Weil • O. Kokshagina
MINES ParisTech, PSL University, Centre for Management Science, 60 Boulevard Saint Michel, 75 272, Paris, Cedex 06, France
e-mail: pascal.le_masson@mines-paristech.fr

© Springer Japan 2015

T. Taura (ed.), *Principia Designae – Pre-Design, Design, and Post-Design*,
DOI 10.1007/978-4-431-54403-6_13

13.1 Introduction

Risk management is often seen as reducing the risk associated to a set of alternatives, be it technological risk or market risk. In *practice*, this raises critical issues since it requires to be able to list the risks and assess their probability of occurrence. Hence great debates on nuclear (risk assessment: the probability is very low but the consequences are so terrible that even low probability can not be neglected) or GMO (how to identify the type of risks), and more generally on "safety first" principle (also called precaution principle), which tends to favor "no action" alternative, since it is often the one implying the least (known) risks. Another issue appears in cases of "high uncertainty" or so called "unk unk", where the level of unknowness is so high that neither the level of risks nore even the list of risks are known. In R&D contexts, this corresponds to "double unknown" situations where neither technologies, nor markets are known. These situations are often considered as just unmanageable and they are left to gamblers and random processes. More *formally*, this approach of risk management actually consists in applying models of *decision making* under uncertainty to select among a set of given designs and probable states of nature the one that maximizes the expected utility. Hence design is supposed to be already done; this approach does not directly consider design as a way to deal with risk. In this paper we would like show the paths opened by a design perspective on risk management.

On the one hand, a design perspective leads to add a new action possibility in the model: to design a new alternative to deal with the probable states of nature. On the other hand, the new alternative design might also "create" new risks, so that a design perspective leads also to model the emergence of new risks as an exogenous "design process". Hence a design perspective raises two issues: can we design an alternative that would lower the risk? Does this new alternative create new risks?

Main results: (1) we show that minimizing known risks consists in designing an alternative whose success is *independent* from all the known risks—this alternative can be considered as a generic technology. (2) we show that the design of this generic technology (or the design of the new independence) depends on the *structure of the unknown*, ie the structure of the space generated by the concept of risk-free alternative. (3) we identify new strategies to deal with risks and show that risk management in a design perspective shifts from dealing with uncertainty to dealing with the unknown.

13.2 Part 1: Setting the Issue : Beyond Decision Making, Generic Concepts to Design low Risk Alternatives

13.2.1 Models for the Interaction Design/Environment: The Limits of Decision Paradigm

The notion of risk characterizes the consequences of an event for an actor. The event is considered as a "stroke of fortune" (and rather misfortune). If one adds some restrictive hypotheses on the structure of risks (these hypotheses will be discussed later), then a risk can be quantified as the product between the probability of the event and the consequences of this event for the actor.

More generally, without the restrictive hypothesis, one can consider risk as the relationship between, on the one hand, a "design space", where design uses, generates and transforms propositions and possibly leads to new artefacts; and, on the other hand, an "external" world, which can or shall *not* be transformed by design (it is "out of reach" of the design, it is *considered* as the states of nature that can't be changed), but which can interact (and even strongly) with the design. The "external world" is "invariant" by design (Hatchuel et al. 2013). But precisely for this reason, it plays a critical role: as an invariant, it strongly configures the design dynamic and strategies.

We can name many examples of such invariants: for instance, weather conditions: a design can be robust or sensitive to weather conditions, it will not influence the weather; norms, standards, design rules, consumer behavior, "production constraints", "process capabilities"... all are examples of such invariants in the external world. Risks are of this sort: for certain (many?) designs, terrorist attack, tsunami, earthquake, or just rain... are constraints of the external world that can't be influenced by these same designs but can influence them.

Two streams of research treat these kinds of constraints from the external world:

1. *In a design perspective*, a large stream of research, in particular in the US, has discussed the relationship between the design and its environment. In the 1960s, Alexander proposed to consider design as the creation of a relationship between a "form" and a "context". Drawing the boarder between form and context *is* exactly the task of the designer, and, for Alexander, it can be considered as a dual process of problem setting and problem solving (Alexander 1964), which can be supported by more or less sophisticated and abstract patterns. Alexander proposes actually a dynamic process that leads to the stabilization of a set of "specifications" that characterize the way the "context" is taken into account in the design space. Alexander actually opens two difference perspectives to deal with robustness, that are deepened in other approaches:

   – On the one hand, the issue is to deal with a fixed set of functional requirements. More generally in many design theories and methods, "functions" or "functional requirements" precisely appear to play the same role than by Alexander: they represent the influence of the context on the design space (see Aximatic Design (AD) (Suh 1990), General Design Theory (GDT) (Yoshikawa 1981)). And the methods and theories deal with robustness by supporting the design of a solution that bot fit with the functional requirements and can support variations around this target. The design deals with a form of local invariance.

- On the other hand, the issue to "identify" the set of requirements that should be addressed by design. Some theories will focus on the way to identify this set. In Coupled Design Process (GDT) (Braha and Reich 2003), the set of "requirements" is built during the process, since at each stage, the closure space of the tentative design leads to integrate new functions in the design. AD and GDT, can also be re-interpreted in this perspective: since the theories actually propose ways to address a given set of functions, as long as the design space follows some properties (Hausdorff space by GDT; axiom of independence between Functional Requirements (FR) and Design Parameters (DP) by AD), it means more generally that these theories offer ways to deal with complete functional spaces that follow these properties. In this case, robustness can be understood as the capacity to design a solution for large functional spaces, that follow specific properties. Or: if the "context" follows certain structural properties, then a design is possible. For instance in AD: if the "context" can be described a set of FR and if the FR and the DP follow the "independence" axiom, then a design is possible. The same for GDT: if the entity space is structured like a Hausdorff space, then a design is possible. This perspective explains that if the "invariant"—ie the context, like FR-DP and their relationship in AD- follows a certain structure, then the design is possible. This also holds for CDP and Infused Design (ID) (Shai and Reich 2004a; Shai and Reich 2004b): CDP explains the effect of closure on the design; ID explain how to design by making use of structures in multiple domains.

To summarize: design theories and methods deal with robustness in two ways: either they address the issue of *local invariance* or they help to *address global invariance by adding specific conditions on the structure of invariance*; but they hardly study the interaction between varied structures of invariance and the design space, ie the effect of certain structures of invariance on the design and conversely the sensitivity of design to certain structures of invariance.

Note that C-K theory can accept an invariance in K (see in particular (Hatchuel and Weil 2007)) and, contrary to other theories, does not make any hypothesis on the structure of invariance. This explains why it was possible to use C-K theory to study how the notion of "face force" emerged from an original "infused design" process in which the structures of two design domains didn't correspond exactly, the "hole" in the two invariant structure being the wellspring of an innovative design process. Hence it seems interesting to use C-K theory to study the consequence of certain invariance structures on design.

2. *Decision-making perspective.* The relationship between design and external environment has long been studied in a completely different stream of research, namely decision-making (Savage 1972; Friedman and Savage 1948; Raïffa 1968; Wald 1950). In decision-making, the question is to find a choice function that helps to identify the design alternative that best fits a set of probable states of nature. Contrary to the design- approaches, the set of states of nature can follow very general structures—it is only supposed to be a probability space-, just like the set of alternatives (which can be a simple list). Note that, even if this approach is not "design-oriented", decision-making largely used in R&D and engineering department to decide on the portfolio of projects to be launched. The "probable states of nature" are the markets and the decision alternatives are the products (or the technologies) to be developed. Another very useful case of application that uses the same framework are Taguchi methods, which actually help to analyse what is the best alternative to meet with varied external contexts.

   Let's analyse the Wald statistical decision-making model (derived from decision theory), one of the most general models of decision-making under uncertainty. The models unfolds as follows:

   - There are states of nature θ in Θ and random variables X_i within R^n with $L(X_1, X_2,... X_n, \theta)$ the likelihood function for $(X_1, X_2... X_n)$ and $\mu(\theta)$, a priori density on θ. This is a representation of the states of nature, the probability of states is subjective and there is model associated to the subjective probability: the random variables model knowledge creation on these same states of nature—the a priori probability density on θ depends on X_i.
   - There is a set D of decisions δ (These are the known alternatives: known technologies, known products...)
   - The relationship between the states of nature and the set of known decisions is modeled as follows: these is a cost function $C(\theta,\delta)$ that associates a certain cost to every pair (state of nature; decision) [this cost function can also be a utility function, see (Savage 1972)]; the action consists in deciding for a certain δ_ι, depending on θ, $\mu(\theta)$ and $\underline{x}_i$ (the results of the sampling operation). Ie one looks for a decision function ψ,

     $$\underline{x} \in R^n \xrightarrow{\psi} \lambda_{\underline{x}}(\delta)$$

     that leads to decide for a certain δ_ι. According to the decision-making theory, the best decision function, the function that minimizes the expected cost function:

$$E(C) = \rho(\mu, \psi) = \int_{R^n \times D \times \Theta} C(\theta, \delta)\lambda_{\underline{x}}(\delta)L(\underline{x}, \theta)\mu(\theta)d\delta d\underline{x}d\theta \qquad (13.1)$$

Hence decision-making models helps to select the design alternative that has the best fit with a set of (subjectively) probable states of nature θ, $\mu(\theta)$ according to a cost function. This alternative is the "most robust" to the uncertainty on the states of nature. In this model, dealing with uncertainty does *not* consist in designing (the set D of alternatives δ_t is closed); it consists in acquiring knowledge to reduce the uncertainty on θ. Risk management is an uncertainty reduction process, not a design process. Still a research project can precisely be financed to reduce some technical uncertainty or, in marketing, a market research project can gain knowledge to reduce uncertainty on consumer behavior.

Hence the decision making perspective is very general and does not depend on invariance structures; but it enables only a limited form of action. Our goal in this paper is to cast this decision making approach into a design paradigm. We will see how this operation helps (1) to identify a new set of concept, namely "generic concept"; (2) some specific features of the design of generic concepts; (3) dynamic models with repeated interactions between "invariants" and related design space.

13.2.2 Casting the Decision Model into a Design Framework: The Logic of Generic Concepts

To make a first step to cast decision making into a design perspective, let's rework the equation of the Waldian model. For sake of simplicity we consider that the sampling is reduced to 0—ie there is no opportunity to gain more knowledge on the states of nature. Suppose that we can design an alternative δ_{n+1} that would be better than all the other alternatives. It is easy to prove that the only property required by δ_{n+1} is:

$$\forall i = 1 \ldots n, \int_{\Theta} C(\theta, \delta_{n+1})\mu(\theta)\, d\theta < \int_{\Theta} C(\theta, \delta_i)\mu(\theta)\, d\theta \qquad (13.2)$$

Or, without simplification:

$$\forall i = 1 \ldots n, \int_{R^n \times \Theta} C(\theta, \delta_{n+1})\lambda_{\underline{x}}(\delta_{n+1})L(\underline{x}, \theta)\mu(\theta)\, d\underline{x}\, d\theta$$
$$< \int_{R^n \times \Theta} C(\theta, \delta_i)\lambda_{\underline{x}}(\delta_i)L(\underline{x}, \theta)\mu(\theta)\, d\underline{x}\, d\theta$$

In a design perspective, this equation actually is *the brief* (in C-K theory: the concept) *of a "robust" design*. This concept still depends on a priori probability μ. Actually such a δ_{n+1} following Eq. (13.2) would be the best for all μ in the domain M:

$$M(\delta_{n+1}) = \left\{ \mu / \int_{\Theta} C(\theta, \delta_{n+1}) \mu(\theta)\, d\theta < \min_{i=1...n} \int_{\Theta} C(\theta, \delta_i) \mu(\theta)\, d\theta \right\}$$

An even more robust solution would be independent of μ. It means that whatever the belief on the states of nature—ie even for states of nature considered extremely low-, the alternative δ_{n+1} is the better. This can be written as:

$$\forall i = 1 \ldots n, \forall \mu, \int_{\Theta} C(\theta, \delta_{n+1}) \mu(\theta) d\theta < \int_{\Theta} C(\theta, \delta_i) \mu(\theta)\, d\theta \qquad (13.3)$$

Example: Raincoat Cap

Let's illustrate what it means on a simple example of decision making situation (see Fig. 13.1). Suppose that the decision maker wants to have a walk and his decision space is D = {d_1: take a cap to protect against the sun; d_1, take a raincoat to protect against the rain}; the states of natures are $\Theta = \{\theta_1$, sunny weather; θ_2, rainy weather}; and subjective probability are, for instance {$\mu(\theta_1)=0{,}51$; $\mu(\theta_2)=1$-$\mu(\theta_1)=0{,}49$}. The utility function is U $N^2 \rightarrow R$, for instance : U(θ_1; d_1) = 100; U(θ_2; d_2) = 10; U(θ_1, d_2) = 100; U(θ_2; d_1) = 10. This situation is usually represented by a decision-hazard tree (see Fig. 13.1). One computes the expected utility associated to each decision. With these data, the decision-maker should choose d_1—and one also understands the fragility of this choice, due to the proximity between $\mu(\theta_1)$ and $\mu(\theta_2)$. This remark usually leads to increase knowledge (e.g. look at weather forecast even if it reduces the walk time).

If we add the hypothesis that the actor can design a new solution, then the dominating solution can be designed as d_3 such that U(θ_1; d_3) = 100; U(θ_2; d_3) = 100. The design of d_3 might lead to a kind of "raincoat-cap".

Note that industrial history is actually full of such design. The graph in Fig. 13.2 illustrates the fact that some technologies are for instance *independent* of

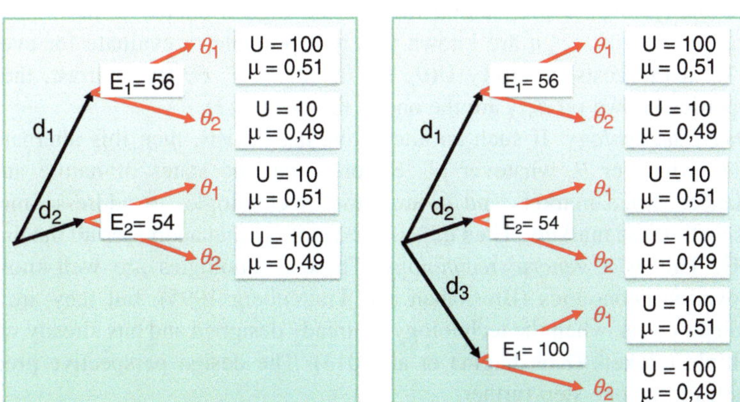

Fig. 13.1 Raincoat-cap example. (**a**) Selection of the solution with the best expected utility; (**b**) design of a solution with dominating expected utility

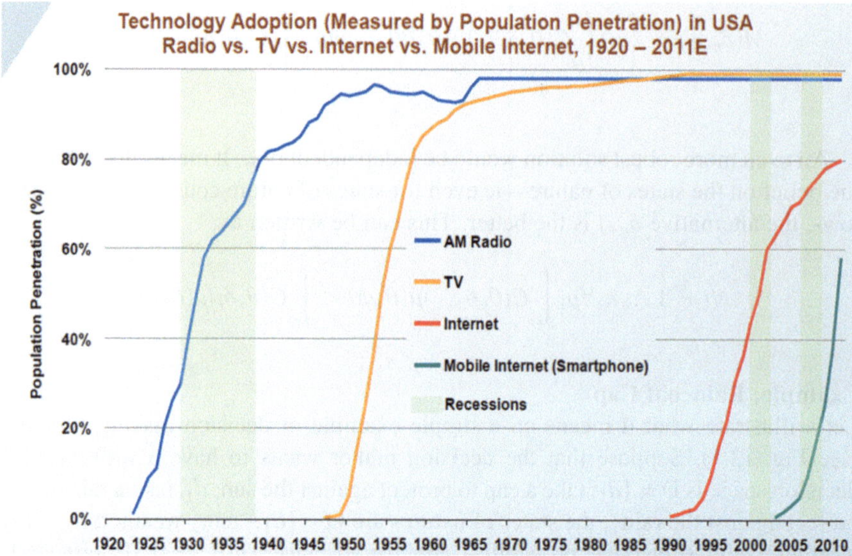

Fig. 13.2 Technology adoption (measured by population penetration in %), in USA, Radio/TV/Internet/Mobile internet, 1920–2011. *Source*: Radio penetration data per Broadcasting & Cable Yearbook 1996, Internet penetration data per World Bank/ITU, Mobile Internet (smartphone) data per Morgan Stanley Research; 3G data per Informa

economics conditions. Even in recessions times, these technologies are successful. As said Jean Schmitt, one famous VC in high tech industry: "some technologies do breakout even in breakdown times"

Comments: Generic Concept and New Relationship "Design Space"/"Context"
Let's underline some properties of the designed alternative:

1. *A conceptual alternative*. The nature of d_{n+1} is very different from all d_i, $i \leq n$: the latter are known alternatives whereas the former is a concept (in the sense of C-K theory: only a proposition that has no logical status). This is precisely because every d_i, $i \leq n$ are known that it is possible to evaluate for every d_i, $i = 1 \ldots n$ the costs or utility $U(\theta_j, d_i)$ for all $\theta \in \Theta$. By contrast, the only properties known on d_{n+1} are the one in the Eq. (13.3).
2. *Generic technology*. If such an alternative d_{n+1} exists, then this alternative is valid "whatever θ, whatever μ". Suppose that the states of nature are, for instance, varied markets and d_i are varied technologies to address—more or less well- these markets. Then d_{n+1} is a technology that addresses all the markets in Θ. This is a *generic technology*. These technologies are well-known in innovation economics (Bresnahan and Trajtenberg 1995); but they are often studied ex post, when the technology is already designed and has already won on multiple markets (Kokshagina et al. 2013). The design perspective precisely enables to go one step further:
3. *Generic concept*. Since strictly speaking, d_{n+1} is not known but is a concept, we call it a *generic concept*. If this concept is designed, it would lead to a generic

technology. *Hence Eq. (13.3) is actually the brief for any generic concept given a set of (subjectively) probable states of nature.*

Let's underline some consequences of this concept for the relationship between design space and "context" and for risk management:

1. *Invariance.* The new design space opened by d_{n+1} did *not* change the states of nature—they are still *invariants.*
2. *Independence.* But this new design space is made *independent* of these states. d_{n+1} *creates* a new *relationship* to the invariants.
3. *Risk management.* In this perspective, risk management does not consist in reducing uncertainty (even if this track remains open and the reduction of uncertainty also changes the design possibilities for d_{n+1}). We have already noticed that this design perspective does actually correspond to the design of generic technologies. But it would also be interesting in a lot of "controversies" situation (see the twentieth century controversies on "smoking" or on "asbestos cancers" or today the debates on potential danger of electromagnetic waves created by wifi and mobile networks): usually these controversies are based on a logic of uncertainty reduction (hence debates on the "proofs" of "cigarette's cancer" or "asbestos's cancer"). In a design perspective, Eq. (13.3) leads to ask for the design of a concept like "as much pleasure as cigarette but independent of any risk of cigarette' cancer" or "as much fire protection as asbestos but independent of any risk of asbestos cancer". Note that it also opens a "design approach" for the "safety first" principle: the safety first principle requires that a technological alternative is *chosen* only if there is no risk; the default is that, with such a formulation, the principle can only be applied to "known" solutions. A design perspective of "safety first principle" actually leads the actor (for instance government, citizen associations...) to ask for the study of concepts that would follow Eq. (13.3).
4. *Creativity and system engineering synthesis.* The design of independence is actually at the root of a lot of engineering efforts. This is precisely what is required by the first axiom of Axiomatic Design or by Taguchi quality principles. It helps also to understand *a specific for of creativity in engineering design synthesis*: engineering design synthesis and creativity are often found contradictory—creativity brings a new dimension or a new technique that doesn't fit with existing systems and the synthesis rather consists in adapting the creative efforts to all system constraints. A generic concept is actually a creative path to deal with systems constraints by becoming independent of them! A generic concept does not add constraints but rather suppress (some of) them.

In a more dynamic perspective, the generic concept leads to study differently the states of nature:

1. *Long term stability.* Being independent of a large set of states of nature, one can also consider that a generic technology is compatible with them. If over time the probability of states evolve (this time in a "natural way", not as a change in the subjective probability), the generic technology will remain dominating. In the case of raincoat-cap, even if global warming increases the probability of sun, the raincoat-cap remains the best solution. In case of technologies and

markets, this means that a generic technology will survive many evolutions on the markets. Dynamically there will be a long term stability of a generic concept.

2. *Expansion in the states of nature*. On the other hand, the new technology will be sensitive to "new" states of nature, ie states of nature that were not in Θ, states of nature that were unknown—and not uncertain. It means that the generic concept actually "opens" a new set of risks that is strictly speaking unknown, that is *not* a combination of already known states, as if "nature" would be designing new states! Hence the design perspective also lead to *introduce a design logic in the regeneration of the states of nature*. For instance one can imagine that the raincoat-cap technology might be sensitive to electromagnetic waves. Hence the set of states of nature should be extended to include states like "electromagnetic storm". *Hence the risk emergence process should be considered as a design process, and even an expansive one.*
3. *Risk regeneration as an expansive design of the states of nature*. Note that this process of "expanding the states of nature" is not a modification of subjective probability, but it actually consists in (re)designing the probability space of the states of nature. The basic ground, the probability space Ω, A, P (where Ω is the sample space, A is the σ-algebra of events and P is a probability measure function—see Kolmogorov axiomatic) is extended to a new Ω'.

To conclude: the design perspective on risk leads to formulate specific concepts, that we call "generic concepts", which are of the form "there is an alternative that has a high utility whatever the states of nature taken in a set of states Θ". Such a generic concept consists in designing an independence relationship with some invariants, namely the elements of Θ. In the next part we will a model of the design of such an independence. Based on this model, we will then study, in the last part, how this new design dynamically interacts with "invariances" that evolve over time.

13.3 Part 2: The Design of Generic Concepts in Matroids and Algebraic Extensions Models

13.3.1 Beyond an Evolutionary Model of Generic Technology Design: The Example of Watt and Boulton Reciprocating Steam Engine

Having identified a generic concept, we are interested in studying the design of such a concept. There are already implicit models of the design of generic technologies. The usual one is an "evolutionary, random" model: a generic technology is a randomly emerging design that is applied progressively on a sequence of applications, and application after application this design appears dominating on a (large) subset of these applications. The story of the steam engine is often told this way: Watt designed a steam engine and progressively many applications were found for it.

Still the evolutionary model, where one "species" progressively adapt to multiple environments, is only one possibility. As shown in (Kokshagina et al. 2013;

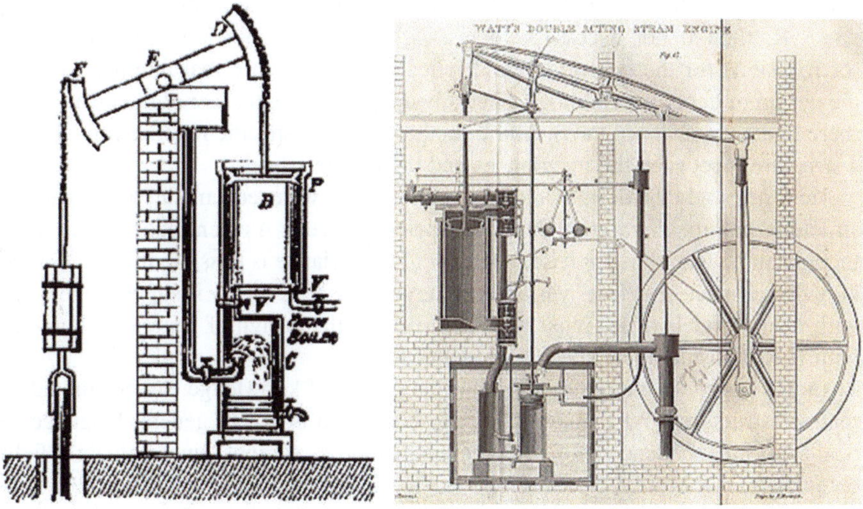

Fig. 13.3 The design of steam engine as a generic technology, by Watt and Boulton. (**a**) 1763 Watt steam engine with separate condensation chamber; (**b**) 1784 Watt & Boulton Double acting steam engine (parallel motion or so-called "reciprocating steam engine")

Kokshagina et al. 2012), this model actually does *not* correspond to how Watt and Boulton historically designed steam engine as a generic technology: based on historical books, Kokshagina et al. show that there was already steam engines, and they were adapted to mining, but not to other uses; hence steam engine in the 1970s was not a generic technology (see illustration in Fig. 13.3). In the 1780s, Boulton asked Watt to work on a generic concept "a steam engine that is compatible with multiple machine tools"; and Watt designed a specific "steam-engine technology" for this concept. Surprisingly enough, it even appears that the key issue was very specifically to design a new way to transmit movement from steam engine, namely a "reciprocating steam engine". Hence a strange paradox of the design of generic technology: it seems to be a "complete" original technology (steam engine), but its design actually focuses on a detail in the complex system (transformation of the movement of the steam engine rod).

This historical case underlines that the evolutionary model might hide more intentional and complex processes, and there might be a variety of processes to design a generic technology.

13.3.2 The Issue: Designing "Whatever Theta"

The issue is to understand the specificities of a design process that designs one independence between the design space and some invariances. The evolutionary model deals with it in a sequential process: a technology d_0 is designed for one θ_0, and then its utility is tested on each of the other elements θ_i of Θ. But suppose that Θ

is made only of linear variations of θ_0, ie every θ_ι is of the form $\theta_\iota = a_i . \theta_0$ where $a_i \in R$; then it not necessary to test d_0 for each θ_ι but it is enough to test (or redesign) for the *operation* "multiplication of θ_ι with a real number". Hence the design of generic technology takes advantage of the *structure* of the invariances (here linear dependence in Θ). And it finally designs a specific relationship—hence a structure—between the invariances and the design space.

Hence to study the design of generic technology, we need a model of knowledge structures and their evolution during the design process. As mentioned earlier, C-K theory can be useful in our case, because: (1) invariance is possible in C-K; (2) the knowledge space is a "free" variable, in the sense as C-K theory is supposed to work with many models of K. What we need to study the design of generic technology with C-K is just to add a specific model of K.

In this paper we choose to consider that pieces of knowledge can be studied as matroid structures. Why matroid? Because matroid is a very general language to deal with independence in many equivalent models (graphs, linear algebra, field extensions,. . .). Moreover, it offers ways to characterize structures and their evolutions (a structure can be characterized by rank, circuits, bases, lattice of flats; the evolution of structures can be modeled with operations of duals, minors, sums, deletion, contraction, extension,. . .). Hence adding matroid structures to K-space (in C-K theory) provides us with powerful analytical tools to follow the transformation of structures during a process of designing a generic technology. Note that in matroid theory, the operations on matroids were mainly used to "analyse" matroids, ie to identify micro-structures into more complex ones. We will use the same tools to rather understand how, during a design process, new structures emerge from given ones.

13.3.3 A Model of C-K with Matroids in K

Matroid structures were introduced by Whitney, in the 1930s, to capture abstractly the essence of (linear) dependence. A matroid is a pair (E, I) consisting of a finite set E and a collection I of subset of E satisfying the following properties: (i) I is non-empty; (ii) every subset of every member of I is also in I (I is hereditary); (iii) if X and Y are in I and $|X| = |Y| + 1$ (the operator $|\ldots|$ designates the number of elements in a set of elements), then there is an element x in X-Y such that $Y \cup \{x\}$ is in I (independence augmentation condition). I are the independent sets of a matroid on E, M(E). There are many forms for matroid (defined on matrices, on algebraic extensions,. . .). In particular, it is very easy to consider the matroid given by a graph: Given a graph G with vertice V(G), the set of vertices of the graph and E(G) the set of edges of the graph. Then let I be the collection of subset of E that do not contain all of the edges of any cycle closed path (or cycle) of G. Then (E, I) is a matroid on G; it is called the cycle matroid of the graph G and is noted M(G).

We use this structure to model the design process associated to the design of generic technologies:

Fig. 13.4 A graph G

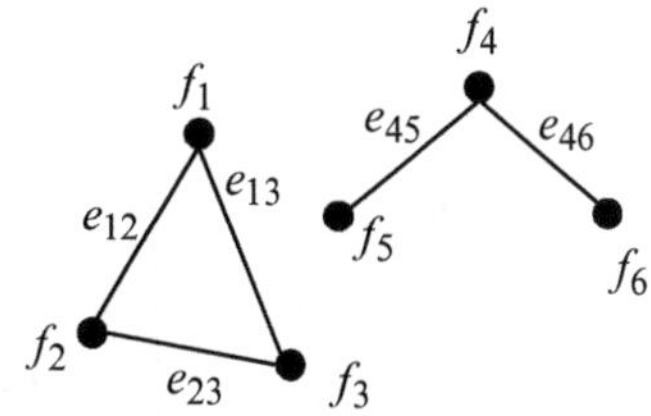

1. *In K*: We consider that K contains a graph G. One interpretation of this graph can be as follows: the vertices are some functions fi; an edge represents a technology to address a pair of functions. A path of edges defines a technology (a combination of technologies) to address all the vertices in the path. The graph G below can be interpreted as a synthesis of the technological know-how of the designer. For instance in the Fig. 13.4 the designer knows how to address $\{f_4; f_5\}$ (with the edge e_{45}); and he knows several solutions to address $\{f_1; f_2; f_3\}$ (e_{12}-e_{23} or e_{13}-e_{23}); he doesn't know any solution to address $\{f_5; f_6\}$ or $\{f_3; f_5\}$. This graph of designer's knowledge is a matroid.

   Note that this matroid can also be characterized by its basis or its cycles. It has a certain rank, which actually corresponds to the size of the largest independent set. In a graph G, we have: rank(G) = |V(G)|-ncc(G) where |V(G)| is the number of vertices in G and ncc(G) is the number of connected components in G. (rank (G) = 4 in the example below). The rank illustrates the most "complex" technologies that can be built by the a user of the K-space: at most the graph G below enables to separately address four independent edges.

   The matroid also has so-called "flats". A flat is a set of elements such that it is impossible to add a new element to it without changing its ranks. One says that a flat is a "closed" set in a matroid. In G below, $\{e_{12}, e_{23}, e_{13}\}$ is a flat and $\{e_{12}, e_{13}\}$ is not a flat. The flat represent stable "substructures" in a matroid.

   We also represent in K the context invariance, representing all the states of nature that the design could be facing. We consider for instance that the designer might need to address any combination of functions taken from the set of functions $F = \{f_1; f_2, \ldots f_n\}$. And we consider (for sake of simplicity) that any market is independent of the other (the market $\{f_1; f_2\}$ is independent of $\{f_1; f_2; f_3;\}$ even if the functions of the former are included in the functions of the latter). This context invariance is also a matroid: it is one of the so-called "uniform matroids" $U_{p,n}$ where the elements are the n-first integers and the independence sets are all subsets of E of size equal or less than p. The contexts can hence be represented by a matroid $U_{n,n}$, in which all the subsets of E are independent. With these two matroids we have the structures of the known, in K space.
2. *In C:* In the model, designing the generic concept consists in designing additional edges in G to address all the markets in $U_{n,n}$. It actually means to design the missing edges so that the graph built on F becomes complete—ie: each vertex is linked to all the others by one single edge. This complete graph is called $K_{n\text{-}1}$ in matroid theory. It means that if $F = V(G)$ and G is complete, then the designer knowledge is a generic technology for $U_{n,n}$ built on F.

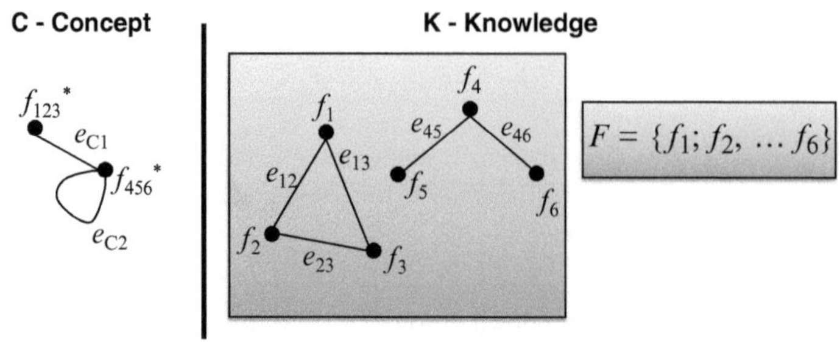

Fig. 13.5 C-K structures with matroids in K

In C, it is possible to represent the graph of all the missing and required edges: from the graph G, one represents all the missing and required edges; one contracts all the edges that exists and are necessary; and one skip all the edges that are known. If $F = \{f_1; f_2, \ldots f_6\}$, this gives the C-graph in Fig. 13.5. Note that this C-graph is also a matroid. It is a connected matroid, and, deleting the loops, it is even a complete graph. Its rank equals the number of connected components in K.

Let's see how the design of the missing edges occurs. There are two different processes to add one single edge, and the matroid model will help us to understand their critical features:

1. *Constant-rank extension.* This concerns the loop-edges in C. the graph G can be "completed" with a new edge following a (single) *extension* process that does *not require changing the rank of the matroid.* It means that the new edge does not add a level of sophistication, a new dimension to the graph. It can be proven (see (Oxley 2011)) that any single extension of this kind corresponds to a flat whose rank will be unchanged by the new edge. Such a flat is called a modular cut. In graph G above, the edge e_{56} linking f_5 and f_6 can be added without changing the rank of the matroid; and it corresponds to the modular cut $\{e_{45}; e_{46}\}$.

   It can be shown that this process can be repeated until it completes all the connected components.

   The constant rank extension is a process to create new *dependent* sets. For instance the new edge e_{56} creates also $\{e_{56}; e_{45}; e_{46}\}$. More generally, the structure of the dependent sets can be measured by the rank of the dual of the graph, also called the corank. If the graph is a spanning tree on n vertices, the rank is n-1 and the corank is zero (no dependent set in the graph). If the spanning tree is completed following the constant-rank extension, its becomes a complete graph K_{n-1}, the rank is kept to n-1 but the corank will reach (n-1)(n-2)/2.

   This description corresponds to the fact that for a set of functions $\{f_4; f_5; f_6;\}$, all the associated markets will be addressed and for the most sophisticated markets involving more than two functions, there will be several technological combinations to reach it. The constant-rank extension hence helps to address all the markets and, simultaneously, to offer multiple technological alternatives to address each markets. This actually corresponds to a very powerful form of robustness.

2. *Rank-increase extension.* The constant rank extension is not enough to address all situations of generic technology design: the issue rises when the set F contains vertices from different connected components in G. It is then required to design a new edge. This new edge e has very different properties as the edge designed by constant-rank extension: every flat of G will change rank when e is added and *no new dependent set is created.* This extension process hence *adds one dimension* to the graph (Oxley 2011). For this reason, this can be assimilated to an expansive partition.

   Such an edge corresponds to the fact that two distinct sets of functions that were not connected at all will now be connected by one single new edge. This opens the possibility for more complex technologies. Note that adding such an edge always means that the new "technology" is *compatible* with all already existing technologies (since it enables all new paths that use the new edge and other existing edges). A good example of such an "edge" is Watt and Boulton "reciprocating movement" that enabled to link the steam engines technologies (one connected component) to the machine tools technologies (another connected component)

The complete design process will (at least) then proceed as follows:

1. Constant-rank extension until all connected components are complete.
2. Connect connected components. There are several possibilities here, since the C-graph is complete and only a spanning tree is required. Suppose that all the C-edge have a certain cost: then the problem consists in finding the spanning tree of minimal weight over a matroid structure. This can be done with the Greedy algorithm (Kruskal algorithm in the case of graphs).
3. The resulting graph in K is now connected but is not complete. It is necessary to identify in C the missing edges and to proceed with anew with a constant rank extension. Some properties of this third step should be underlined:

   a. The new edges that will be created are necessarily associated to a flat that contains at least one edge resulting from an expansion process. In this sense all the new edges are now the consequence of the expansive edges.
   b. But on the other hand, each new edge of the third step creates a new circuit that includes an expansive edge. This means that it also creates *substitute* for the expansive edge.

Note that this process is not deterministic and keep the "generative" aspect of design: for instance there are many possibilities for the spanning tree, depending on the weight that the designer will put on all the C-edges. For instance the weight can be based on the estimated cost of the technology development; but it can also be linked to the expected difficulty to make this new edge compatible with already existing edges in the two distinct connected components,... Many other weight systems are possible—note that whatever the weights chosen, the matroid structure of the C-graphs warranties convergence of Greedy algorithm.

We say that the process will *at least* go through the three steps: actually additional steps are possible. For instance, the designer might himself add a new "function" to be addressed—this is far from unusual: it would just means that the designers add a constraint. Of course this new function should be then be handled to keep independence.

13.3.4 Main Properties for the Design of Generic Technologies

The model with matroid helps to understand *critical properties on the design of generic technology (GT)*: GT results from two "genericity building" operations: G1 (in step 2) connects by preserving (necessary) past connections; and G2 (in steps 1 and 3) completes the graph, ie creates new dependent structures in G based on the first new expansive edge. These two processes explain apparent paradoxes in the design of GT:

1. Because of G1: A generic technology includes both the longest cycle, hence the "most constrained" one, *and* a local property (just one missing edge). This corresponds to the logic of genericity in steam engine, where genericity was made by working on reciprocating movement to link the technologies of steam engine with the technologies of machine tools. More generally, the design of generic technology consists in designing to creates new compatibilities between "islands" of disconnected technologies.
2. Because of G2: Every new edge is a consequence of the first expansive edge (hence a function of it) but, in the end, every new edge is also a possible substitute for the initial expansive edge. Hence a generic technology also encloses variances and alternatives. It is not one single solution but rather a set of interdependent solutions.
3. *Discussing the evolutionary model of GT design.* We can easily represent in the matroid model the evolutionary process mentioned above for the design of GT: it consists in building new edges for each single new market. It is interesting to note that this kind of random, evolutionary process can certainly help to complete a connected graph but will hardly manage to create the edge that connect two separate connected components: of course one market might require to link to vertices from disconnected components, but this market would hence allow for a technology that is not necessarily compatible with all already existing technologies in the connected component. This only possible for such a connection is a market that needs all the function of the two disconnected components. *Hence a random, evolutionary process is quite unlikely to lead to a generic technology; it should at least be guided by the requirement to connect disconnected components.*

The model also enlightens some aspects of *"independence"* in generic technologies:

1. The design of GT actually establishes a certain structure between designed world and environment. Here this independence is actually modeled as a complete graph K_{n-1}. One can underline that in terms of matroids, this structure has interesting properties: K_{n-1} and $U_{n,n}$ have the same rank; one can even notice that their lattice of flat are isomorph (see (Oxley 2011)); but they have very different corank, 0 zero for $U_{n,n}$ and (n-1)(n-2)/2 for K_{n-1} and hence very different dependent sets. *This means that a generic technology appears as a set of compatible, more or less substitutable technologies.* In a sense the knowledge set of an engineering department should embody a generic technology!
2. Note that the design process actually works on several structures: two structures in K, modeling what is known; and also a structure in C. The latter is interesting: this is *the structure of the unknown!* This structure represents all the "holes" in K and their interdependences. The design consists in building this structure and "extracting" some edges from it to make them become knowledge. It can be noted that the C-structure follows a matroid structure which enables a relatively easy process to identify a spanning tree in the C-graph.

13.4 Part 3: Dynamic Models of the Interaction Designed World/Environment

Based on the model above, we can now study the dynamic interaction between a designed world and some "constraints".

13.4.1 Dynamic Model

As already noticed in part 1, the logic of generic technology leads to a new logic to follow the dynamics of context. As long as a new "context" is actually a dependent set in the structure of external structures, then this context is taken into account by the generic technology. Hence the only change in context that would lead to a new design are context that include a property that is independent from the past *F*. It can be a new function f_{n+1}.

In a dynamic process, this kind of extension might lead to several forms:

1. If the successive sets of risks F(t) are included one after the other (for each t, $F(t) \subset F(t+1)$), then the process leads at each time t to a complete matroid K(t) with $K(t) \subset F(t+1)$
2. But there is another possibility: suppose that the set F(t + 1) contains a new function f_{t+1} but does not contain all previous functions. Then in C, edges that are known and not necessary are deleted. By decreasing the requests for

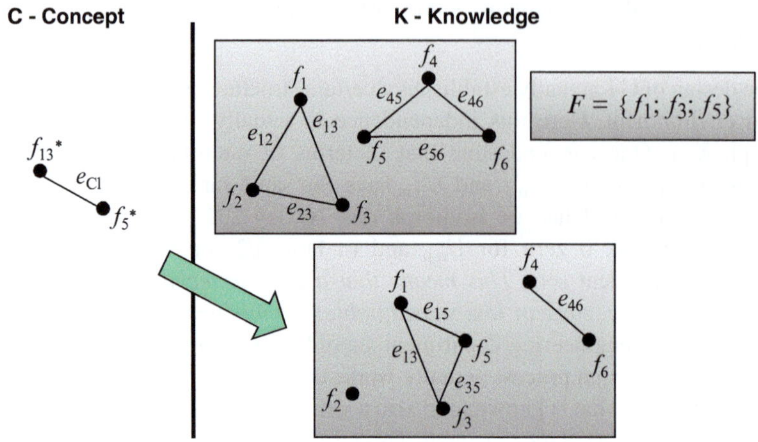

Fig. 13.6 Dynamic models of interaction designed world/context

compatibility, this simplifies the task to link the new function to the past one. Back in K, this leads to a new connected component associated to the set F(t + 1) but also to other connected components associated to the functions that were connected earlier but are not more necessary for F(t + 1) (see Fig. 13.6)

This second process actually leads to the emergence of complex but separate connected components in K.

13.4.2 Using the Model to Interpret Some Dynamic Situations

1. A first situation to analyze is the *"imperfect design"* of generic technology: in part 2, we made the hypothesis that the designer would be able to address all the edges required. Suppose now that there is a budget limit. Then it is possible that the budget is not sufficient to create a spanning tree for the C-graph. Or the designer will select a subset of F to get only a partial generacity. We say that the design of GT is imperfect. Interestingly enough, these behaviors can be represented in the model above. They all fall in the second situation, either because the set F is restricted in advance, or because the new edges to be designed can only address a subset of F., due to cost reasons. It explains how over time, one can have the stabilization of a connected subgraph without relation with active parametres of F.
2. The dynamic model also leads to come back to evolutionary models in a richer way: instead of considering a selection process based on one given state of nature, one can represent the evolutions of probable states of nature, ie as subsets of F. In this extended evolutionary model, one can find *partial genericity*

changing over time and separate connected subgraphs that can suddenly appear useful if, at a certain time t + q, F(t + q) suddenly reuses some of the "old" functions and hence the connected component becomes relevant again.

13.4.3 One Application: Evolutionary Models in Biology

The dynamic model of the design of generic technologies might be relevant for some technological evolutions; but one can underline that actually this model might even be relevant for models of evolutionary process in biology.

Suppose that one represent a species as a designer who master certain "technologies" to address external conditions; suppose that these conditions evolve over time but the species never completely adapt to one context but keep a "robust" strategy by adapting to several contexts. The species actually acts as a designer of a (partially) generic technology. What does our model help to explain:

- The model is coherent with stasis and punctuated equilibria: if the context evolves inside the set of alternatives addressed by the species, the "technologies" are not changed; the species evolve only when an "original" function, out of the robustness scope, appears.
- The model is coherent with *exaptation situations* (Gould 1987): exaptation is a situation where an organ (a technology) is developed to adapt to a situation—the panda's thumb was developed to go up and run in trees at a time where panda were smaller and lived in a dense forest context-; then this function is less needed but the organ remains and the species go on evolving, keeping the original trait—the panda doesn't need to go up and run in trees but changes in his environment lead him step by step to eat bambous-; and finally the function "holding a branch" become important again but this time to eat bamboo: the panda's thumb becomes important again.

Hence our dynamic model could account for contemporary forms of complex, evolutionary processes in biology!

13.4.4 One Application: Evolutionary Models in Biology

We already underlined (part 2) how our model could help to design generic technology (constant-rank extensions and rank-increase extensions). Some lessons can also be learnt from the dynamic model: it appears that the critical issue is the emergence of the new function. In part 1 we considered that this new function could be interpreted as a form of "expansive design" in nature. In a risk management perspective, it is also important to manage the emergence of the new function. It consists in launching "exploratory projects" that help to extend the list of "risks" (or F in our model).

The dynamic model helps to understand why such an exploration can just work! Apparently, such an exploration is very difficult, since "nature" doesn't speak in advance and the new function can be everywhere! Still the dynamic model shows what should be explored in such an exploration: (1) the risks that are *not* in F. In a sense it restricts the exploration to brand new functions, it leads to better characterize F and its structure to understand what is out of F and its structure; (2) the risks are impacting the known technologies, usually in a negative way since it proves that it is not robust to a new F.

Note that it explains why this kind of exploration was called "crazy concept" exploration: it consists in exploring a concept that is "out" of the usual set of risks (or market opportunity) and that can not be addressed with available knowledge!

13.5 Conclusion

In this paper we study generic concepts, ie concepts for artefacts that would be valid for a large domain, ie for a large set of external conditions. We have shown that these concepts can be derived from decision making theory and are of the form given by Eq. (13.3). Equation (13.3) means that such a generic concept establishes an "independence" with the set of possible contexts. We have then analyzed the design based on a generic concept. We used a model derived from C-K theory with matroids in K-space. We showed that the design process combines two different operations, constant-rank extension and rank-increase extension and these properties explain critical properties of generic technologies as well as the limits of evolutionary processes to design generic technologies. Finally we study dynamic model for the design of generic technology and show that these dynamic models account for "imperfect design" and could open new perspective for the study of classical "risk management" in biology; the study of evolutionary processes.

References

Alexander C (1964) Notes on the Synthesis of Form, 15th printing, 1999th edn. Harvard University Press, Cambridge

Braha D, Reich Y (2003) Topologial structures for modelling engineering design processes. Res Eng Des 14(4):185–199

Bresnahan TF, Trajtenberg M (1995) General purpose technologies: engines of growth? J Econom 65(1):83–108

Friedman M, Savage LJ (1948) The utility analysis of choices involving risk. J Polit Econ 56(4):279–304

Gould SJ (1987) The Panda's thumb of technology. Nat Hist 96(1):14–23

Hatchuel A, Weil B (2007) Design as forcing: deepening the foundations of C-K theory. In: International conference on engineering design, Paris, p 12

Hatchuel A, Weil B, Le Masson P (2013) Towards an ontology of design: lessons from C-K Design theory and Forcing. Res Eng Des 24(2):147–163

Kokshagina O, Le Masson P, Weil B, Cogez P (2012) Platform emergence in double unknown: common challenge strategy. In: R&D management conference, Grenoble, p 25
Kokshagina O, Le Masson P, Weil B (2013) How design theories enable the design of generic technologies: notion of generic concepts and genericity building operators. In: International conference on engineering design. ICED'13, Séoul
Oxley J (2011) Matroid theory. In: Cohen R et al (ed) Oxford graduate texts in mathematics, 2nd edn. Oxford University Press, Oxford
Raïffa H (1968) Decision analysis. Addison-Wesley, Reading
Savage LJ (1972) The foundations of statistics, 2nd edn. Dover, New York (1st edn.: 1954)
Shai O, Reich Y (2004a) Infused design: I theory. Res Eng Des 15(2):93–107
Shai O, Reich Y (2004b) Infused design: II practice. Res Eng Des 15(2):108–121
Suh NP (1990) Principles of design. Oxford University Press, New York
Wald A (1950) Statistical decision functions. Wiley, New York
Yoshikawa H (1981) General design theory and a CAD system. In: Sata T, Warman E (eds) Man–machine communication in CAD/CAM, proceedings of the IFIP WG5.2-5.3 working conference 1980 (Tokyo), Amsterdam, p 35–57

Chapter 14
Computational Schema as a Facilitator for Crowdsourcing in a "Social-Motive" Model of Design

Rivka Oxman

Abstract In this paper we propose and introduce parametric design as a potential methodological basis for a *shared computational representation schema* that can support *crowdsourcing design* in a 'social motive' model of design. A 'social motive' model of design is multi-phase (Pre-design, Design, and Post-design) model of design that offers a new perception for obtaining and exploiting *experiential feedback of society*. We first introduce the theory and concepts of *Crowdsourcing*. In the following sections we propose a theoretical schema that demonstrates how information flow is shared in a holistic and compound model in various phases of design. Our digital design model and the conceptual structure represents computational design processes of informed design model; informing design about per-formative and physical environmental conditions as well as the experience and wisdom of the crowd which can be easily mapped to the pre-design; conceptual design; and post-design. Finally we illustrate, demonstrate and discuss how by exploiting the concept of the *parametric schema* as a common representational formalism, the wisdom of the crowd can potentially be integrated to inform processes of digital design generation, adaptation and change in the various phases of design in order to improve design.

Keywords Coding • Collective intelligence • Crowdsourcing design • Digital design • Informed design • Parametric design • Parametric schema • Scripting • Social motive • Social network

---

R. Oxman (✉)
Faculty of Architecture and Town Planning, Technion Israel, Haifa, Israel
e-mail: rivkao@gmail.com; rivkao@tx.technion.ac.il

© Springer Japan 2015

T. Taura (ed.), *Principia Designae – Pre-Design, Design, and Post-Design*,
DOI 10.1007/978-4-431-54403-6_14

14.1 Introduction

14.1.1 *The "Social Motive": Integrating Social Knowledge in Design*

Since the middle of the last century extensive research has been carried out attempting to develop unified models of design. Most procedural models present design as a linear series of sequential phases consisting of *analysis, synthesis and evaluation* (inter alia: Asimov 1962; Archer 1963; Lawson 2006). Early *computational models of design* have also focused on the iterative nature of the design process by attempting to accommodate the interdependencies of design where modifications in design are required and the cyclical processes are therefore iterative (Maver 1970).

A new perspective in the development of models of design has recently been proposed by Toshiharu Taura (Kobe University) and Yukari Nagai (JAIST) (Taura 2014a, b) emphasizing the relationships among an extended number of design phases. Their proposed phases are described as: Pre-Design, Design, and Post-Design.

This extended sequential model offers a new perception of the potential of exploring methods for obtaining and exploiting *experiential feedback of society* that they have generally defined as "social motive". According to this approach, "in the Pre-Design phase, for example, the explicit or implicit social motive is identified and translated into an explicit requirement or specification" that may result in improvement in the design process. Furthermore, lessons learned from Post-design experience of "design in use" can contribute to new perspectives and perceptions to improve future design. Therefore, elicitation of social response and experiential knowledge, developing simulation techniques, evaluation and appraisal may constitute a promising method to improve design.

14.1.2 *Crowdsourcing Design in a Multi-phase Model*

In consideration of the potential research and design attributes of the extended stage model, we are currently faced with two methodological challenges: the development of a *medium for the acquisition of social knowledge* in each of the three phases; and the development of a *computational medium for the integration and adaptation* of this knowledge within the three inter-related phases. We present in the following current on-going research *a computational medium for social knowledge acquisition, design integration and design adaptation.*

In an initial research phase we have been exploring the potential of, and approaches to, *crowdsourcing design* by adapting concepts such as *the wisdom of the crowd.* Today, despite the broad interest in adopting crowdsourcing in design, there is a lack of disciplinary conceptual foundations, formal guidelines, and supportive web techniques that can effectively facilitate the potential of crowdsourcing in design.

In this paper we propose and introduce parametric design as a potential methodological basis for a *shared representational schema* that can support crowdsourcing design in an extended model. We illustrate, demonstrate and discuss how by exploiting the concept of the *Parametric Schema* as a common representational formalism, the wisdom of the crowd can potentially be integrated to inform processes of parametric adaptation and change in the various phases of design in order to improve design.

We conclude the paper by discussing and proposing a future agenda to facilitate crowdsourcing in design, proposing for the development of *Parametric Design Schema* as shared representation in order to provide operative methods in the Pre-Design, Design and Post-Design phases.

14.2 Crowdsourcing Design

14.2.1 Introduction

The theory of *Crowdsourcing* shares certain common concepts with the fields of *Collective Intelligence* (Lévy 1997) and *Design Collaboration.* Today there is rapidly growing interest in research related to the application of emerging techniques of collective intelligence (Maher et al. 2010) and crowdsourcing in design (Oxman and Gu 2012). In order to understand the basic principles of crowdsourcing there is a need to initially determine "how the power of the many can be leveraged to accomplish feats that were once the responsibility of a specialized few" (Howe 2008). According to Jeff Howe the crowd has the ability to contribute uniquely to the production of creative and innovative ideas replacing traditional forms of thinking, making and doing.

While crowdsourcing has already proven successful in knowledge derivation in various fields, the current lack of a theoretical basis and conceptual understanding must be addressed in order to facilitate the potential of crowdsourcing in design. This requires a program of research exploration and experimentation that *studies crowdsourcing methods in explicit reference to design media and design techniques* (Oxman and Gu 2012).

As an initial basis for understanding the motivation and the power of crowdsourcing it is most useful to compare seminal foundational concepts in related fields. *Collective Intelligence* has been one of the significant models for crowdsourcing. While psychological studies of collective intelligence are based on the observation of human behaviour, biological studies are frequently inspired by the observation of swarm behaviours in nature. For example, the computational modelling of the biological behaviour classified as "swarm computing" has been influenced by the observation of such instinctive biological behaviour.

According to Lévy, one of the foundational figures in this field, *Collective Intelligence* is a "universally distributed intelligence" and as such promotes the "universality of intelligence" (Lévy 1997). According to Lévy, collective intelligence can be characterized by three concepts:

1. Universal distribution of intelligence
2. Constant enhancement
3. Coordination in real-time and the mobilization of skill.

The phenomenon of the "wisdom of the crowd" (Surowiecki 2004), is supported by web-based technologies and interactive communications media. In parallel to the development of web communication, the model of the crowd represents an efficient model for use in decentralized models and decentralized participation. Crowdsourcing (Surowiecki 2004) is an emerging concept demonstrating the breaking of traditional centralized models of collaboration. The difference between crowdsourcing and the traditional *outsourcing* is that issues that need to be solved can be open and distributed to a body of unknown potential contributors rather than specific recognized collaborators.

Crowdsourcing is now becoming an accepted model for "online, distributed problem-solving and production" (Brabham 2008). The model has potential for solving issues in different fields exploiting a massive crowd of online users in fields like art, music and photography and has been recently presented and discussed in the context of design (Oxman and Gu 2012; Oxman and Oxman 2014).

14.2.2 From Collaborative Design to Crowdsourcing Design

Collaborative design is relatively well understood and has been theoretically formulated in design research. Collaborative design implies well-organized teamwork and negotiation. Design collaboration depends on designers within the discipline as well as design experts across the discipline. Teamwork and negotiation processes are based on shared domain knowledge and disciplinary expertise. However beyond the traditional conditions of collaboration, the emerging complexity of global projects; the need for new models of global practices; and the influence of a global economy have created demands to find solutions for extended needs, changing scales, and new modes of design (Lahti et al. 2004).

While decision-making and collaborative problem solving process in crowdsourcing can be guided by statistical results, design as a cognitive activity is characterized by the uniqueness of its thinking processes (Cross 2011). Furthermore, in contradiction to the statistical or optimal outcome that can be associated with the non-hierarchical collective social intelligence of crowds, design is a task domain that focuses on unique and specific representational codes and operative skills. As such, collective intelligence in design must address specific knowledge and skill-media that are based on accepted representational codes, methods and processes as well as upon unique bodies of disciplinary knowledge.

To summarize, crowdsourcing technology, based on social media technology, may potentially support the contributions of individuals in a decentralized digital environment to share their knowledge and wisdom. In comparison to collaborative design, crowdsourcing can attract a large number of unknown potential participants representing different levels of domain-specific knowledge, interdisciplinary knowledge and expertise who may be interested in, and useful for, design crowdsourcing.

14.2.3 Scenarios in Crowdsourcing Design

In order to explore basic issues for applying crowdsourcing in design we have considered the following three potential media-related scenarios (Oxman and Gu 2012): crowdsourcing design in open-source social networking; crowdsourcing design through open-source modeling; and crowdsourcing design in open-source generic prototyping.

These are described below.

- **Open source 'social networking'**
  A social network is a social structure of individuals or organizations termed 'nodes', which are linked by specific types of relationships, interests, beliefs, knowledge, etc.

  Social networking technologies (i.e. Blogging, Facebook, Wiki, Twitter, etc.) may be considered as Open-source scenarios in crowdsourcing design.

  An example is the Wikipedia platform (http://www.wikipedia.com). Wikipedia is a web-based encyclopedia project supported by the non-profit Wikimedia Foundation. The information in Wikipedia, is supplied by employing traditional media of text and visual communication: text articles, images and links to other sources of information are usually coordinated by ad hoc *editors* that are interested in specific *topics*.
- **Open source 'modeling'**
  The open-source modeling scenario is related to 3D modeling in online environments. In the case of a specific design problem scenario, open-source modeling could support *brainstorming sessions* instigated by modeling and visualization of 3D models in addition to verbal input.

  "User-generated content" producing 3D models in web-based online environments is fundamentally different from the production of 'user-generated content' in Wikipedia. The sharing of design media and technology supporting the collaborative manipulation of design seems to be an essential ingredient here. For example, the concept of online design collaboration and participation in virtual worlds has been explored by an *open design group*, composed of individuals from various disciplines, interested in exploring the application of an open-source platform to share ideas, issues and production processes in architecture and urban design

This example follows a Wikipedia concept and is called, the *Studio Wikitecture* (*Studio Wikitecture*—http://studiowikitecture.wordpress.com). The studio enables a self-organizing network of collaborators using textual and graphical media to develop an experimental environment, sharing ideas, and exploring the validity of design manipulations through crowdsourcing.

- **Open source 'coding and scripting environments'**
  Finally, the open coding and scripting environments for the creation of design prototypes can be developed as an environment for supporting shared collaborative crowdsourcing in design similar to the application of *Arduino* as an open-source electronic prototyping platform for creating interactive objects or environments (http://www.arduino.cc).

  We can observe these three basic formats as potential foundations of crowd sharing in design: the communality of the subject, or (design) problem in the Wiki; the communality of the modeling environment; and the communality of the software medium as a shared framework and collective basis for the open-source creation of crowd knowledge. Finally, in this work, we propose that a shared computational scripting environment can function to integrate the advantages of these three basic formats.

14.2.4 The Social Motive

The "social motive" (Taura 2014a, b) *adds a general temporal (or stage-based) framework to crowdsourcing-design.* The temporal framework further defines *the nature of knowledge to be solicited.* The stage-based framework addresses the need for the *acquisition* of specific forms of social experience, experiential design knowledge related to successes and failures, the sharing of unique non-standard situations, questions, and set of problems and solutions. The sharpened temporal perspective of the "social motive" aids in making specific the well-defined role of *informed relationships* among design variables, solution space options, etc. as they relate specifically to the extended number of design phases.

Understanding the role of information flow in a period of social and cultural change (Oxman 2013) as well as the *changes of information flow* in each of the three stages is critical for research and development of crowdsourcing design.

The potential development of a theoretical framework for Crowdsourcing in Design in Pre-Design; Conceptual Design and Post Design raises the following issues:

- **Informed processes of design**
  We have proposed a theoretical schema that demonstrates how information flow is shared in a holistic and compound model in various phases of design. Our conceptual structure represents computational design processes of informed design model; *informing* design about per-formative and physical environmental conditions as well as the experience and wisdom of the crowd which can be easily mapped to the Pre-design; Conceptual design; and Post-design as suggested by Toshiharu Taura and Yukari Nagai (Taura 2014a, b).

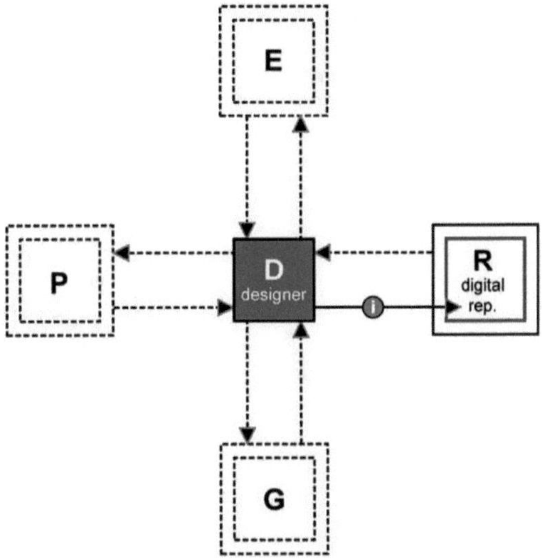

Fig. 14.1 Informed process schema of design. *D* designer, *R* representation, *G* generation, *E* evaluation, *P* performance (Oxman 2006)

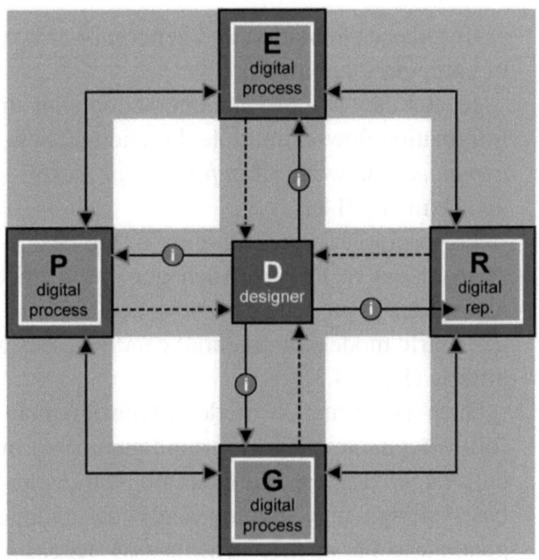

Fig. 14.2 Integrated compound network of informed processes of design (Oxman 2006)

The proposed model contains four basic components that represent four classes of the traditional design activities: representation; generation; evaluation and performance (Fig. 14.1).

- **Integrated compound model of design**

  The proposed framework demonstrates a compound model of design includes: Formation model, Evaluation model, Performance-based model and Generative model (Fig. 14.2).

  Furthermore, our proposed compound model of design has opened up a way to link informed processes of analysis, and generative and evaluation processes (Oxman 2006) by various computational methods of analysis, evaluation,

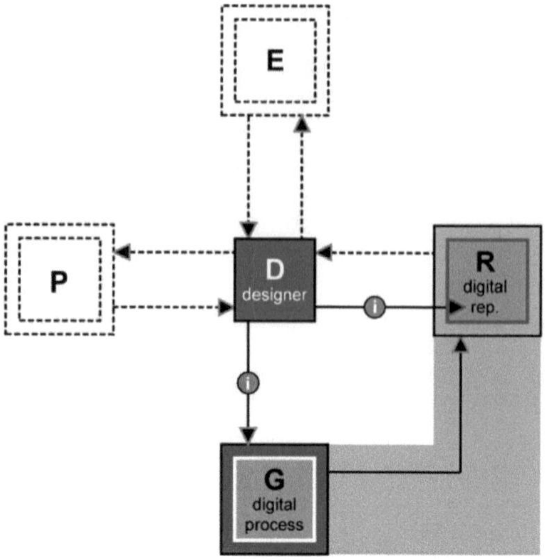

Fig. 14.3 Generative process of design. *R* representation, *D* designer, *G* generation (Oxman 2006)

performance simulation and generation that can be integrated into various forms of computational media.

Ideally they will provide interaction with any activity module with the data and information flow in multiple directions. These may be thought of as an *integrated compound* network of *informed processes* of design (Oxman 2006; Oxman and Oxman 2014).

For example: crowdsourcing in the design phase can be supported by generative model of design. Such generative models of design employ generative algorithms to produce generic schema. These in turn can produce specific visual geometric models of potential candidate design forms that may meet the desired criteria (Fig. 14.3).

Such performance model of design may also include, among others, the following parameters: environmental performance, financial cost, spatial, social, cultural, ecological and technological perspectives. Furthermore, performance-based design may employ analytical techniques and simulations (P) that can generate a parametric schema (G) according to parametric variables to map desired performance (Fig. 14.3). These in turn can produce geometric model of desired design form (Oxman 2006).

An interesting possibility in the Pre-design phase consideration of crowd information (such as environmental conditions, climate consideration, traffic, external forces, social behavior and environmental properties, etc.) through crowdsourcing can be supported by a model of performance-based design activating *simulation process* that can be used to evaluate both physical and human behavior before materialization and final solutions (Figs. 14.4, 14.8, 14.9, and 14.10).

– **Shared computational design representation**

Various approaches and media of crowdsourcing design such as open-source social networks generally lack the methodological ability to support informed design processes, beyond textual or visual critique. In order to integrate methods

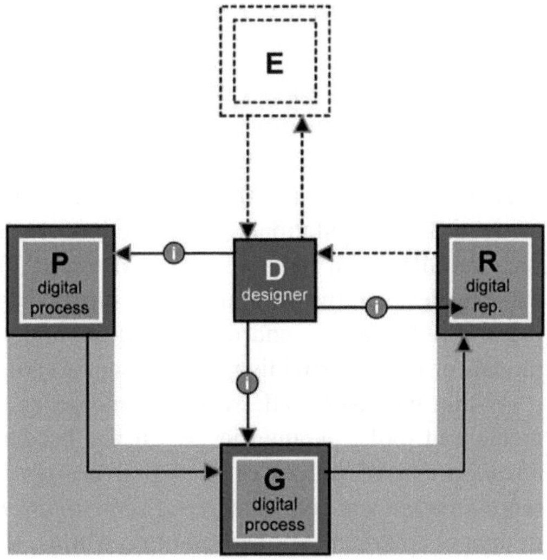

Fig. 14.4 Performance-based process of design. *R* representation, *D* designer, *P* performance, *G* generation (Oxman 2006)

for informing processes within open-source social networks, computational design methods would require the capability to support a *shared computational design representational system.*

Traditional representations in design are focused upon the representation of the design object. Visual representations are non-explicit with respect to presenting the structural logic behind the construction and the design development of the object under design. In order to achieve this provision of information there is a need to explicate the logic that is embedded in process-models of design. This is a key characteristic for the achievement of successful processes of crowdsourcing design: *the knowledge and constructive logic must be made explicit within the computational system of representation.*

One promising method to achieve this level of richly informed socially-based design is the exploitation of a generic approach to *parametric design*. Representational formalism of parametric design can potentially be shared among designers in crowdsourcing design. This approach offers different methods of representation such as algorithmic languages, scripting and coding as a basis for forms of process-based representation in crowdsourcing design. In parametric design, computational algorithms play an important role in generating, manipulating and evolving the design solutions.

14.3 Parametric Schema for Crowdsourcing Design

In the following section we illustrate, demonstrate and discuss how by exploiting the concept of the *parametric design schema* as a common shared representational environment through which the wisdom of the crowd can be integrated in a multi-phase design model illustrating theses ideas in Pre-design, Conceptual design and Post-design.

14.3.1 Introducing Parametric Design

We have seen that early general models of design have focused on the iterative nature of the design process. These recursive models have concentrated on the iterative relationships among analysis, synthesis and evaluation (Asimov 1962; Archer 1963; Lawson 2006). CAD models have later introduced visualized geometrical 3D computational models to help visualize form manipulation and modification after each iterative stage (Maver 1970). One of the difficulties of using these models was the difficulty of the support of interactivity. A major problem was the ability to interact and understand the underlying representational logic of 3D models of design providing transparency to allow shared modifications.

Within the last decade with the emergence of new parametric design environments and tools, parametric design has become a medium in the evolution and advancement of processes of digital design. Parametric design focuses on the representation and control of *associative relationships between objects*. As a result, designers communicate between two worlds, the abstract, based on mathematical and parametric rule-based algorithmic space, mapping a "tangible-realistic space which interacts with the needs of people, communities of different scales" (Ottchen 2009; Yu et al. 2013).

What is distinctive about parametric design is that beyond the focus upon the object of design, parametric design has now moved designers *to define and characterize the formalization of parametric design as holistic processes explicating their key concepts*. Terms and concepts such as *parametric schema*, *generative processes*, *algorithmic scripting and coding* are becoming an important body of thought and interactive methods for shared representation.

14.3.2 The Concept of Parametric Schema

As compared to the explicit representation of the geometry of the form itself, associative and parametric geometry present a logical structure behind algorithmic procedures of design processes. The manipulation of visual design representations is supported by algorithmic representation of its internal logic using representational languages of scripting and coding (Oxman 2009; Oxman and Oxman 2014; Jabi 2013; Woodbury et al. 2007).

The logic of a classical generic schema can be demonstrated through the example of Architectural Buildings such as Greek Temple, Egyptian Temple, Gothic Church etc. (see Fig. 14.5). Today, algorithmic representation of generic design schema of contemporary practice cab be consistent, coherent, and can be applied in contemporary design to produce modification and variations of design (Figs. 14.6 and 14.7).

Fig. 14.5 Classical type: a Greek Temple

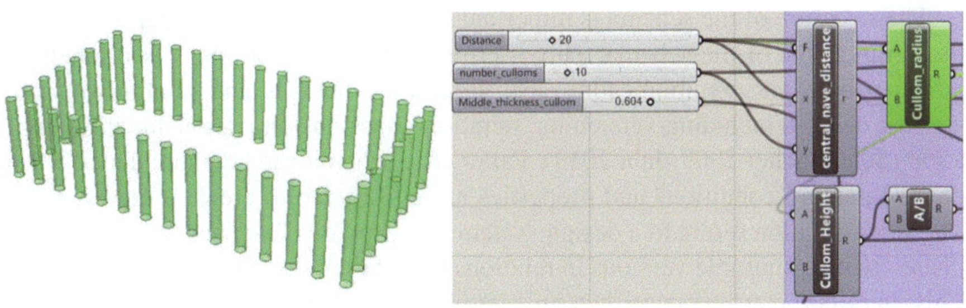

Fig. 14.6 Scripting generic schema: integrated environment. Grasshoper-scripting (*left*) and Rhino-modeling (*right*) (developed by ILLia Musizuk)

14.3.3 Parametric Schema as a Facilitator in Crowdsourcing Design

The design of a structured *parametric schema* is a key concept for the achievement of shared representation in crowdsourcing design. A parametric schema is a collection of structured relationships between functions, parameters, and a geometric model while the form itself is modeled by the structured parameters (Oxman and Oxman 2014; Davis et al. 2011). The legibility of the parametric representation and its capability of eventually becoming an accepted shared representation are central

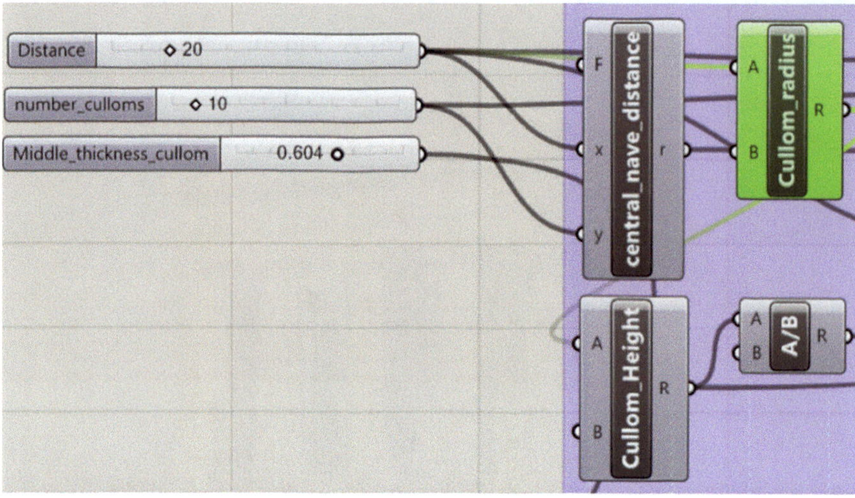

Fig. 14.7 Parametric structuring of flow of information including modification sliders in Grasshoper (developed by ILLia Musizuk)

factors in determining how effectively the potential crowd can interact with the parametric schema.

The structure of the schema is important in making sense of the relationships of the components of the schema. This is critical for a collaborative environment in which people are communicating and suggesting changes and modifications of specific designs including versioning, iteration, mass-customization and differentiation (Schumacher 2009; Jabi 2013). Parametric schema can be manipulated when new contextual conditions and properties are considered to create a new version. The term iteration refers to a design system that provides feedback at any iterative phase creating multiple versions in relations to different sets of desired parameters. This, in effect, is what occurs in a process of crowd-sourced design. Finally, mass-customization supports the creation of customized prototypes by uniquely and easily changing the parameters of each individual element.

14.4 Parametric Crowdsourcing in a "Social-Motive" Model of Design

In the following section we illustrate an attempt to characterize a "social motive session" demonstrating the role of a Parametric Crowdsourcing Design Schema as a shared representational medium that can support social communication in the three phases of design. We demonstrate the potential of a *virtual visualization environment* to support crowdsourcing parametric design illustrating how the generic

Fig. 14.8 Virtual visualization environment to support crowdsourcing design by exchange and modification of scripting files (developed by Itamar Eitan and Mor Herman)

scheme can be dynamically modified and manipulated in various phases of design in a social environment.

- **Virtual visualization environment for crowdsourcing parametric design**
  In the examples illustrated, we demonstrate how parameters and sliders in a parametric system can be used to dynamically modify form in real-time (Fig. 14.8) within a shared social environment for design, in this case, similar virtual visualization environment could also be supported in a web-based social medium for design (Fig. 14.9).
- **Modification of the parametric in pre-design phase**
  In the pre-design phase of crowd-sourcing design the emphasis is placed upon the modeling of performance conditions within the methodological framework and format. In the pre-design phase consideration of crowd information (such as environmental conditions, climate consideration, traffic, external forces, social behavior and environmental properties, etc.) can be supported by a simulation model of performance-based design activating simulation process evaluating both physical and human behavior before the materialization and final solutions.

  Furthermore, the performance-based design model (Fig. 14.4) may include among others the following parameters: environmental performance, financial cost, spatial, social, cultural, ecological and technological perspectives and may employ analytical techniques and simulations (P) that can generate a parametric form-finding schema (G) according to parametric variables to map desired performance (Fig. 14.10). In generation, this would occur primarily in the Design phase. However, it is also characteristic of the schema modifications that would occur (or be socially proposed) in the Post-design phase.

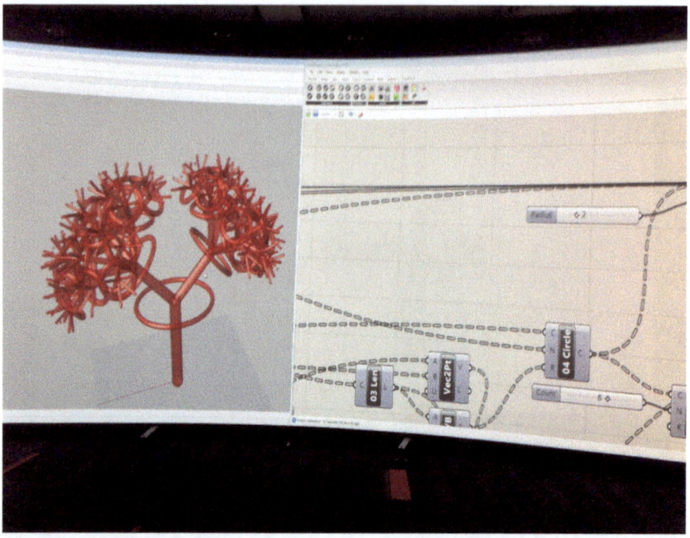

Fig. 14.9 Typical Grasshoper environment supporting exchange and modification of scripting files in virtual visualization environment (developed by Itamar Eitan and Mor Herman)

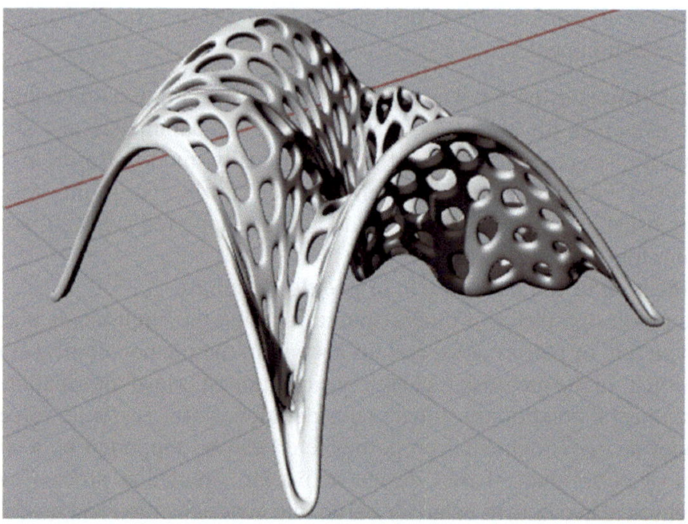

Fig. 14.10 Performance based schema of form-finding process (developed by ILLia Musizuk)

It is interesting to observe that the three phases each become *methodologically defined* according to their relationship to the specific processes of refinement and adaptation that are applied to the parametric schema. The emphasis shifts in the three phases from parametric performance *specification* to *generation* to *modification* (Fig. 14.11).

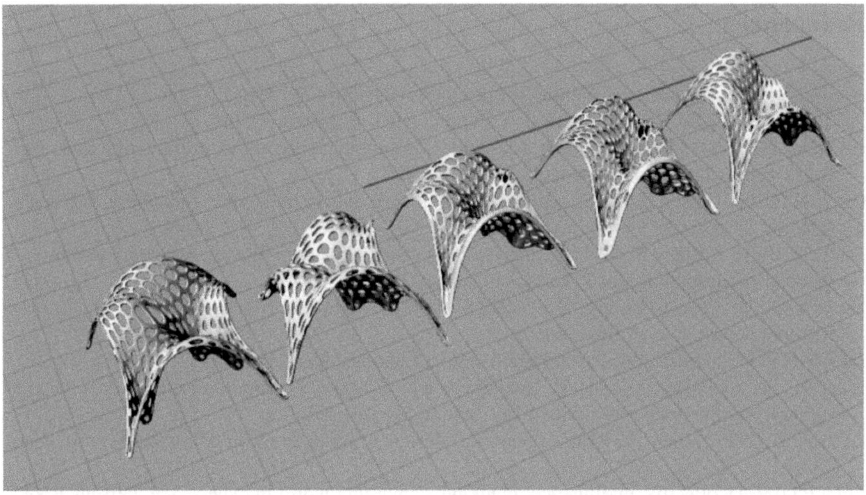

Fig. 14.11 Virtual visualization of parametric crowdsourcing of modified design (developed by ILLia Musizuk)

14.5 Summary and Conclusions

Parametric models can be visualized, informed by geometrical modifications as well as analysis procedures of contextual conditions such as environmental and structural properties, material behaviour, materialization and fabrication requirements, and constructional logic. As Schumacher has stated, the parametric is intrinsic to all architectural design operations: "Systematic, adaptive variation and continuous differentiation concern all architectural design tasks from urbanism to the level of tectonic details.... on all scales" (Schumacher 2009). Thus parametric schema provide both for the social acquisition of knowledge as well as its integration in various processes of refinement and adaptation in design.

We have demonstrated the relevance and the need for the development of a theoretical foundation for an extended model of "social motive" within the framework of crowdsourcing design. Design feedback and knowledge-acquisition in Pre-design in Post-design can, in turn, provide a useful framework for the crowdsourcing of design.

Future development of crowdsourcing design technology integrating shared parametric representation for the acquisition of expertise and experiential knowledge appears to be a promising paradigm for the development of the "Social-Motive" Model of Design.

Acknowledgement I would like to thank the students who participated in my graduate course on "Parametric Design" (Faculty of Architecture and Town Planning, Technion IIT, 2013) for their enthusiasm, curiosity, and their innovative contributions.

References

Archer J (1963) A systematic method for designers. Design 174(188):172
Asimov M (1962) Introduction to design. Prentice-Hall, Englewood Cliffs
Brabham DC (2008) Crowdsourcing as a model for problem solving: an introduction and cases, convergence. Int J Res New Media Tech 14:75–90
Cross N (2011) Design thinking: understanding how designers think and work. Berg Publishers, Oxford
Davis D, Burry J, Burry M (2011) Understanding visual scripts: Improving collaboration through modular programming. Int J Architect Comput 9(4):361–375
Howe J (2008) Crowdsourcing: how the power of the crowd is driving the future of business. Random House, Crown Business
Jabi W (2013) Parametric design for architecture. Laurence King, London
Lahti H, Seitamaa-Hakkarainen P, Hakkarainen K (2004) Collaboration patterns in computer supported collaborative designing. Des Stud 25(4):351–371
Lawson B (2006) How designers think: the design process demystified. Architectural Press, Oxford
Lévy P (1997) Collective intelligence: mankind's emerging world in cyberspace. Perseus Books, Cambridge
Maher ML, Paulini M, Murty P (2010) Scaling up: from Individual design to collaborative design to collective design. In: Proceedings of the DCC'10, Stuttgart, pp 581–600
Maver TW (1970) A theory of architectural design in which the role of the computer is identified. Build Sci 4(4):199–207
Ottchen C (2009) The future of information modeling and the end of theory: less is limited; more is different. Architect Des 79(2):22–27
Oxman R (2006) Theory and design in the first digital age. Des Stud 27(3):229–265
Oxman R (2009) Performative design—a performance-model of digital architectural design. Environ Plann Plann Des 36(6):1026–1037
Oxman R (2013) Perspectives on design innovation research in a period of cultural change: technology-driven informed design innovation. In: Editorial board essay of IJDCI on perspectives on design creativity and innovation research, vol 1, no. 1, p 16
Oxman R, Gu N (2012) Crowdsourcing: theoretical framework, computational environments and design scenarios. In: Proceedings of the 30th international conference on education and research in computer aided architectural design in Europe (ECAADE 2012) on digital physicality | physical digitality, Prague
Oxman R, Oxman R (2014) Parametrics: the design of multiplicities. In: Oxman R, Oxman R (eds) Theories of the digital in architecture. Routledge, Taylor and Francis Group, London and New York
Schumacher P (2009) Parametric patterns. In: Mark G (ed) Patterns of architecture, vol 79, AD architctural design. Wiley, UK, pp 28–41
Surowiecki J (2004) The wisdom of crowds: why the many are smarter than the few and how collective wisdom shapes business, economies, societies and nations. Doubleday Anchor, New York
Taura T (2014a) Motive of design: roles of pre- and post-design in highly advanced products. In: Chakrabarti A, Blessing LTM (eds) An anthology of theories and models of design. Springer, London
Taura T (2014b) Preface. In: Taura T (ed) Principia designae: pre-design, design, and post-design: social motive for the highly advanced technological society, Springer, Japan
Woodbury R, Aish R, Kilian A (2007) Some patterns for parametric modeling. In: Proceedings of the ACADIA 2007, Halifax
Yu R, Gero J, Gu N (2013) Impact of using rule algorithms on designers' behavior in a parametric design environment: preliminary result from a pilot study. In: Proceedings of the CAAD Futures 2013, Shanghai

Index

© Springer Japan 2015
T. Taura (ed.), *Principia Designae – Pre-Design, Design, and Post-Design*,
DOI 10.1007/978-4-431-54403-6

MIX
Papier aus verantwortungsvollen Quellen
Paper from responsible sources
FSC® C105338

If you have any concerns about our products,
you can contact us on
ProductSafety@springernature.com

In case Publisher is established outside the EU,
the EU authorized representative is:
Springer Nature Customer Service Center GmbH
Europaplatz 3, 69115 Heidelberg, Germany

Printed by Libri Plureos GmbH
in Hamburg, Germany